TEN PATTERNS THAT EXPLAIN THE UNIVERSE

Published by arrangement with
UniPress Books Ltd, London, UK, by the MIT Press

COMMISSIONING EDITOR: Kate Shanahan
PROJECT MANAGER: Kate Duffy
DESIGN AND ART DIRECTION: Wayne Blades
ILLUSTRATOR: Richard Palmer
PICTURE RESEARCHER: Natalia Price-Cabrera

ISBN: 978-0-262-54286-9

Library of Congress Control Number: 2021931084

Printed in China

The MIT Press
Massachusetts Institute of Technology
Cambridge, Massachusetts 02142
http://mitpress.mit.edu

Brian Clegg

TEN PATTERNS THAT EXPLAIN THE UNIVERSE

The MIT Press
Cambridge, Massachusetts

CONTENTS

INTRODUCTION

We understand the world around us through patterns. These are not necessarily patterns in the visual sense, but rather occurrences that have some sort of regularity, distributed according to consistent rules. It would be impossible to cope with the world if we didn't have patterns. It would mean that every time we encountered an object, we would need to learn how to use it anew. Instead, we build up patterns—mental models of reality—that inform us of how to deal with, say, an apple or a light switch, so that we don't have to start from scratch each and every time.

Sometimes our need to find patterns leads us astray. For instance, when we jump at a shadow, we are responding to the pattern of a predator, and it is better to get pattern recognition wrong, than it is not to recognize the danger of a predator when it is truly present. Similarly, superstitions are based on assuming the existence of a pattern that doesn't occur. We aren't alone in having superstitious beliefs; for example, pigeons have been shown to exhibit superstitious behavior. If they coincidentally perform a movement a number of times before being fed, they can start doing the same movement when they are hungry, expecting that it will produce food. Prejudice, too, is the result of a misapplication of a pattern, extrapolating from a single example to a wider group. So, for example, if a member of a particular group performs a terrorist act, all members of that group can then become suspect.

However, the fact that patterns sometimes go wrong should not get in the way of their fundamental role in helping us to understand the universe. And science—arguably one of humanity's greatest achievements—is all about finding patterns. When we speak, for example, of natural laws, such as Newton's laws of motion, we are describing an observed pattern in the way aspects of the universe behave. If there weren't such patterns, our environment would be incomprehensible. Each time we observed something it would behave in a different way. There would be no laws, no science, no technology. The universe would be a realm of chaos. But thankfully, for reasons we don't entirely understand, the universe is set in its pattern-structured ways.

The Cosmic Microwave Background

In this book we are going explore ten of the most significant patterns that help us to understand how the universe works. The first is the cosmic microwave background, radiation that comes from everywhere in space and that was set in motion when the universe was less than one million years old. This has subtle variations in intensity that create a pattern revealing how the very earliest structures in the universe were formed.

Minkowski Diagrams

Next, we discover a diagram that shows how two of the most familiar aspects of reality—space and time—are not truly separate, but intertwined. Arising from Albert Einstein's special theory of relativity, the diagrams developed by Minkowski uncover the pattern of interaction between space and time, which results in mysterious effects as a moving object comes closer to the speed of light.

Particle Trail Patterns

A third type of pattern is the shower of new particles produced in a particle accelerator, such as the Large Hadron Collider at CERN, in Switzerland. Particle accelerators are sledgehammers to crack very tiny nuts. They accelerate electrically charged particles—protons in the case of the Large Hadron Collider—faster and faster until they come close to the speed of light, then smash beams of these high-speed particles into each other. Such is the energy of the collision that much of it is converted into matter, according to Einstein's famous equation $E=mc2$. It is in the pattern of the resulting new particles that discoveries, such as the Higgs boson, have been found.

Feynman Diagrams

The next pattern we encounter is also involved in the interaction of tiny particles, in this case also including electrons and photons of light. But, unlike the tangled mess of the patterns produced by a particle accelerator, these are clear, informative patterns called Feynman diagrams. Devised by one of the world's greatest physicists, these elegant representations illustrate how light and matter particles interact with each other, and can be used to simplify the complex calculations required to take in all the possible outcomes of such an interaction.

The Periodic Table

More recognizable for many people is the pattern that dominates chemistry—the periodic table of the elements. Whether it's the appearance of its classic tiles on the show *Breaking Bad* or the table's ubiquitous appearance on chemistry lab walls, this pattern is a familiar sight. But underlying that structure of rectangles is a far more important configuration: the electron patterns on the outside of atoms that determine how every chemical reaction occurs, from the myriad responses that make our bodies function to a familiar reaction, such as rusting.

Weather Patterns

Usually, patterns help us to understand what is happening and to predict the future—and there can be no better example incorporating both the value and dangers of patterns than those that occur in the weather system. Weather has many familiar patterns, yet historically has been almost impossible to predict. Since the 1980s we have become much better at this, due to our understanding of the special kinds of pattern that are known as mathematical chaos.

The pattern in the sky called the Milky Way is now known to be our home galaxy, containing around one billion stars.

Number Lines

Weather gives us an example of surprisingly simple patterns that can produce complex outcomes. Another such example is the number line. The idea of a sequence of numbers along a line, such as the divisions on a ruler, may seem trivial, but it forms the basis of arithmetic and nearly all the mathematical operations most of us make as a daily occurrence. The number line is a pattern that extends regularly into our everyday lives.

Cladograms

A less familiar pattern is one that extends life into the past. Charles Darwin drew a tree of descent in his seminal work *On the Origin of Species* (1859), but the specific structure that is now most used by

biologists and paleontologists is a cladogram. This is a type of information tree that shows when species split off from their common ancestors, driven by genetic data, giving our best picture of the pattern of evolution.

DNA Double Helix

Cladograms reflect the way that biology has moved from establishing the relationship between organisms from their appearance to using genetics. This new capability came from the discovery of one of the fundamental patterns of life: the structure of the complex molecule called DNA. Along the length of a DNA molecule, a pattern of chemical compounds spells out information just as bits do in a computer.

Symmetries

The DNA molecule has a kind of symmetry, as it is designed to be split down the middle, with one half able to specify exactly what is in the other half. Symmetry seems to be the most fundamental pattern in the universe, underlying much of physics from the conservation of energy to the types of particle that make up everything. It seems appropriate, then, that our final pattern is that of symmetry.

As we have seen, patterns give us the means to have a better understanding of how the universe works, from the very small to the very large. These ten patterns, in their different ways, provide a fascinating picture of reality.

1
THE COSMIC MICROWAVE BACKGROUND

Dr. Robert **Wilson** (left) 1936–
Dr. Arno **Penzias** (right) 1933–

ECHOES OF THE UNIVERSE

Just 380,000 years after the Big Bang—around 13.8 billion years in our past—the universe became transparent. Prior to this, matter in the universe was mostly ionized, each particle electrically charged and ready to absorb light. Once the matter became atoms, that light was free to stream across the universe and it has been doing so ever since. As the universe expanded and cooled, the energy of the photons reduced to become weak, all-pervasive radiation in the microwave region. Discovered by accident in 1964 by Robert Wilson and Arno Penzias, by 1992 the cosmic microwave background had been plotted by NASA's COBE satellite, showing tiny variations in the "temperature" of the background radiation. This radiation has been described as the "echo of the Big Bang," but more accurately it is the fetal scan of a baby universe. This is the pattern of the cosmos.

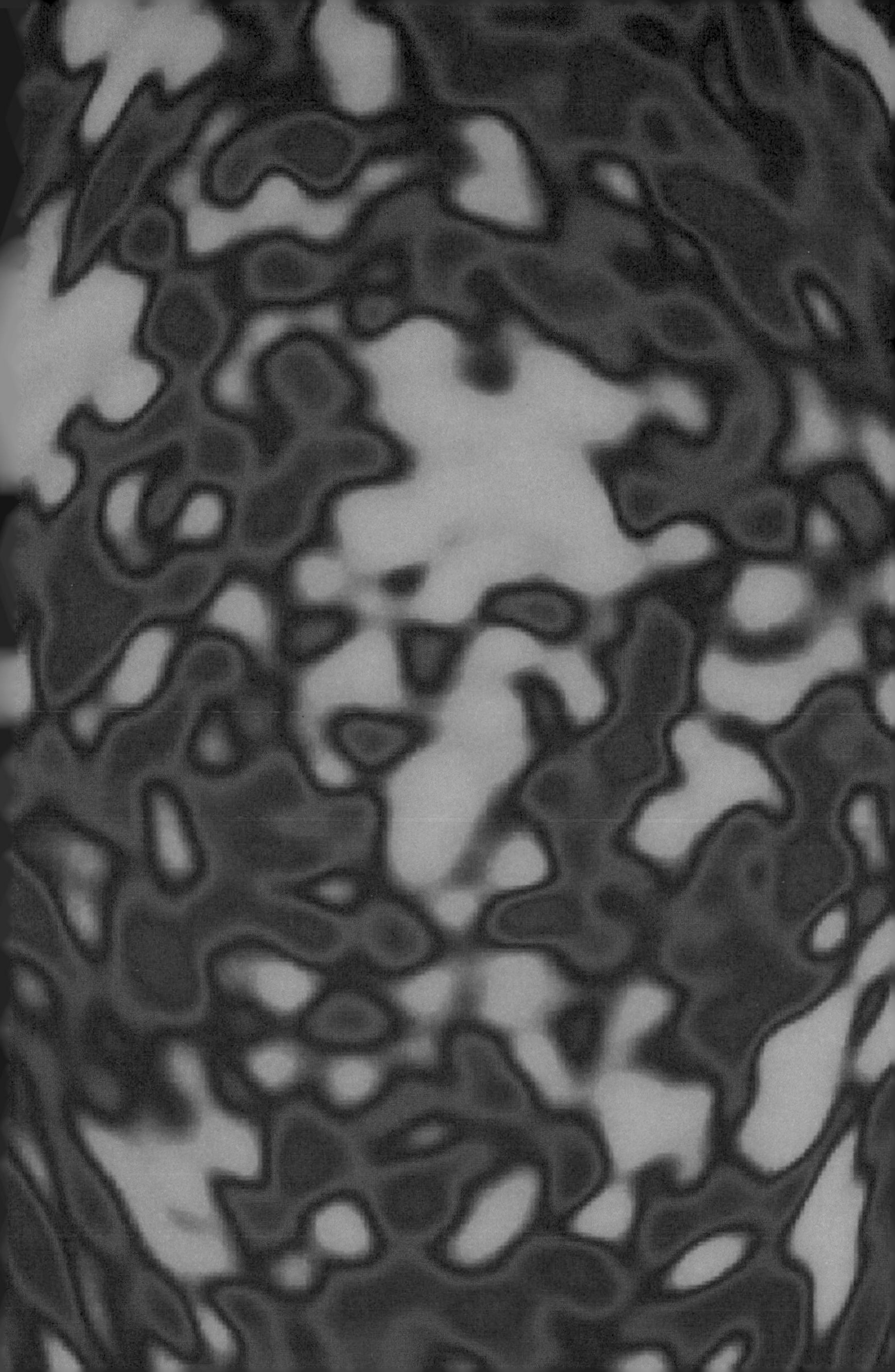

THE EXPANDING UNIVERSE

As the universe continued to expand and cool following the Big Bang, the energy of the photons of light that made up these gamma rays reduced to become faint, omnipresent radiation in the microwave region. Microwaves may be more familiar as the heating mechanism in a form of oven, but they are part of a wide spectrum of light or "electromagnetic radiation" that runs all the way from radio, through microwaves, infrared, visible light, ultraviolet, X-rays, and gamma rays. It's all the same stuff, but as we go up the spectrum, the light becomes increasingly energetic.

This transformation of the light that first flowed across the universe 13.8 billion years ago from ultra-powerful gamma rays to faint microwaves is a direct result of the expansion that the universe has continued to undergo ever since. From the very beginning,

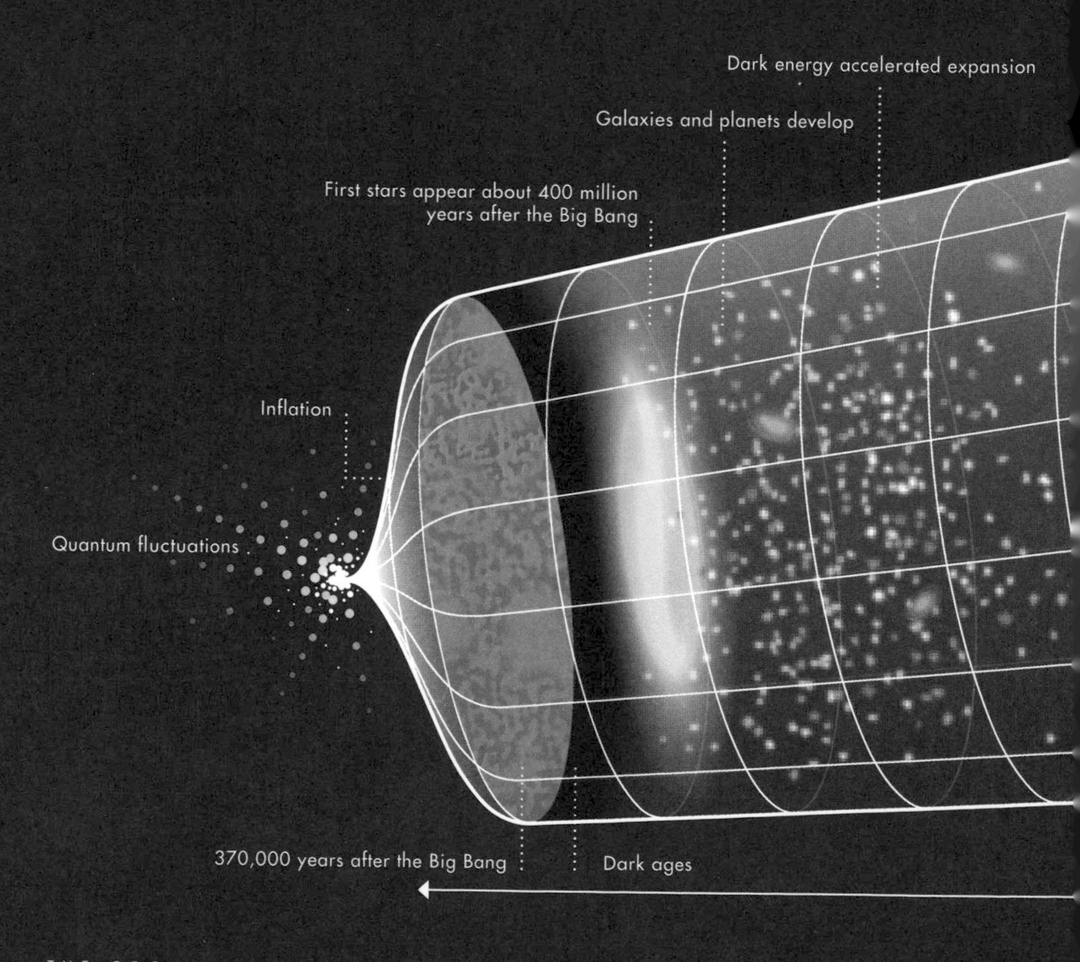

the universe has been expanding—and that expansion is now accelerating, driven by a mysterious force known as dark energy, which is yet to be understood.

When the universe expands, it's not just a case of objects in the universe moving away from each other, the way, for example, an explosion will send bits of matter flying in all directions. It is space itself that is expanding. As a result, if we think of light being a wave, the wavelength of the light increases with the expansion. For visible light, this means moving from blue toward red, while the much shorter wavelength gamma rays would have shifted down through X-rays and ultraviolet, visible light and infrared, all the way to microwaves.

The evolution of the expanding universe

With timescale running from the Big Bang at the left to the present at the right, we can see the evolution of the universe. Over time, the pattern that we see still in the cosmic microwave background provided the template for the development of stars and galaxies.

EXPLORING LIGHT

Another way to look at light is as a stream of particles called photons. With photons, the color of a beam of light corresponds to the energy of the photons that make it up. Gamma ray photons have the highest energy, but this energy is reduced as space expands—like a boxer pulling a punch. So, by now what were initially gamma ray photons have become microwave photons.

By 1992 the cosmic microwave background—the pattern of this light from the early years of the universe—had been plotted by the first of three major satellites, each with a greater ability to uncover detail, showing tiny variations in what is referred to as the "temperature" of the background radiation. In the everyday sense, temperature is a measure of the energy in the atoms of matter, whether we are examining air, or liquids, or solid bodies. When scientists refer to the temperature of light, they are referring to the energy in the photons.

We can't measure the temperature of photons using a thermometer in the usual way—instead this is a reference to the temperature of a particular type of object called a black body, which gives off radiation with photons of this energy. We're all familiar with objects glowing with light when they get hot enough, but in practice even cold objects give off light. We don't see this, as it is low enough in energy not to be visible to our eyes. The temperature of the cosmic microwave background radiation is around 2.7 K on the Kelvin scale, the scientific temperature scale that starts at absolute zero. In more familiar terms, 2.7 K is about -454.81°F (-270.45°C). What we are seeing in this cosmic background radiation is the pattern of the universe.

FIRST DETECTION

The original discovery of this ubiquitous space radiation was something of a fluke. Two physicists working for the US telecoms research company Bell Labs at the Holmdel site in New Jersey found some unexpected interference. The site was the home of an early receiver for signals from space. It had been designed to pick up data from a crude, experimental satellite communications program called Project Echo, which used metallized reflective balloons to bounce signals from one

Temperatures are measured in kelvins (K) on the absolute Kelvin scale. As the temperature of an object increases, the changing spectrum of light it produces makes its glow pass from red to yellow, to white and blue.

point on Earth to another. Following this project, the "horn antenna" as it was known was used on several projects, notably by the pioneering radio astronomers, Robert Wilson and Arno Penzias.

As the name suggests, radio astronomy relies on picking up radio waves, which like visible light are typically produced in space by stars. Wilson and Penzias were attempting to find radio signals emanating from a gas cloud that was thought to exist outside our Milky Way galaxy, but rather than receive a signal from a particular location, they found a background hiss that seemed to come from every direction in space.

The most natural possibility for such a signal was local interference that overwhelmed the antenna's directionality, but even so, some directional variation in this would be expected. However, that was not the case here. Whatever direction the site's bulky horn was pointed in, it picked up the same level of signal. Penzias and Wilson assumed this meant that there was a problem with the receiver itself, as early radio telescopes were often troubled with radio noise produced by their own electronic systems.

After a long process of elimination, the wiring and electronics all tested fine, but the pair discovered that the opening of the horn was covered in pigeon droppings ("white dielectric material" as they coyly described it when writing up their research). It appears that a pair of pigeons had decided that the horn made a good nesting site. Penzias and Wilson arranged for the birds to be sent in the company's internal mail to another Bell Labs site 40 miles (64 km) away, but these were clearly homing pigeons. Within days, they were back at the antenna again. As a last resort, the birds were then shot and the horn was cleaned out, but sadly the pigeons lost their life to no avail. The background hiss was still there, still seeming to come from every direction.

With one of the happy coincidences that often seem to be involved in scientific discoveries, Penzias happened to mention his problem with interference to another radio astronomer, Bernie Burke, when discussing an entirely different topic. Burke put Penzias on to Princeton physicist Robert Dicke, who Burke had heard was working on a topic associated with the Big Bang that might in some way be connected.

GAMOW'S PREDICTION:
In 1953, George Gamow predicted that the cosmic background radiation would have a temperature of around 7 K, very close to the actual value.

Over the years, several theorists—notably the Soviet-American physicist George Gamow—had discussed the possibility that light that was first set in motion when the universe became transparent should still be crossing space. Although this was initially high-energy gamma rays, the expansion of the universe would have lowered the energy, leaving microwaves still traversing the universe. Dicke had set up a small team, trying to detect this background microwave radiation with low-quality, ex-military equipment, but had been unable to pick it out from the general background radio noise that is always present on Earth.

"THE MOST IMPORTANT THING ACCOMPLISHED BY THE DISCOVERY… WAS TO FORCE ALL OF US TO TAKE SERIOUSLY THE IDEA THAT THERE *WAS* AN EARLY UNIVERSE. STEVEN WEINBERG

The New Jersey-based Holmdel antenna had limited value in satellite communications, but made a huge breakthrough in detecting the cosmic microwave background.

Burke called up Dicke after his discussion with Penzias. Dicke realized, with some chagrin, that Penzias and Wilson had scooped him in discovering what would become known as cosmic microwave background radiation. (Although we tend to compartmentalize microwaves and radio from each other because they often have different uses, microwaves are just the subset of radio waves which have a wavelength of between about 1 meter and 1 millimeter.) As a result, Penzias and Wilson (though, perhaps surprisingly, not Dicke, who was responsible for much of the early theory) were awarded the 1978 Nobel Prize in Physics.

FLUCTUATIONS IN THE RADIATION

The defining aspect of the background radiation that was discovered at Holmdel was its uniformity. It initially appeared to be the same from all directions, which though important is not what would usually be thought of as an informative pattern. However, when studied in detail, the radiation was discovered to have very small variations in intensity. It was these fluctuations that would form the pattern that produced the first picture of the early structures of the universe forming. These variations would not be detectable until a satellite could be used to pick up the radiation, away from earthly interference and the distorting effects of water vapor in the atmosphere. The first satellite observatory to successfully do this was known as COBE.

COBE was a satellite more completely known as the Cosmic Background Explorer, launched by NASA in 1989 (it should have been sooner, but the deployment was delayed by the *Challenger* space shuttle disaster in 1986). It was COBE that would produce the first of the iconic elliptical maps, which are projections of a three-dimensional view out in all directions onto a flat surface. This is, in effect, the inverse of the map projections used on maps of the Earth to show the whole planet laid out on a flat sheet of paper (see pages 46–48).

The variation producing that distinctive pattern is the result of very slight differences in the intensity of the microwave radiation. The detail looks quite dramatic, but it is important to realize that there is just a 1 in 100,000 difference between the brightest and dimmest of the intensities shown. In fact, in the very first COBE images shared with the world, the satellite really wasn't sensitive enough to pick out all but the biggest variations—the majority of the pattern that looked so impressive from these early COBE results was actually caused by temperature variations in the microwave detector on the satellite. As first presented, the image was more than a little deceptive.

Since COBE, however, we have had two further generations of satellite that have been able to produce better and better images, eliminating most of the error and showing vast amounts of finer detail. The next to enter the field was another NASA satellite, WMAP (the Wilkinson Microwave Anisotropy Probe), which operated from 2001 to 2010. ("Anisotropy" here simply means that the satellite is

The plot of variations in the microwave background detected by the COBE satellite went through a number of processes to produce a usable image. At the top is an initial combination of the data at two key frequencies. The central image removes variation caused by the movement of the solar system, while the bottom picture also removes the impact of radiation from the Milky Way galaxy.

measuring variations in the background radiation.) Unlike COBE, which was in a conventional orbit around 547 miles (880 km) above the Earth's surface, WMAP was positioned in a tight orbit around a location known as the Earth/Sun L2 point.

The L2 is one of the so-called Lagrangian points, where the gravitational pull from the Earth and the Sun balance each other out. Consequently, a satellite that is sent there will stay in place in a fixed position with respect to the Earth as it passes around the Sun. The L2 point is located on the far side of the Earth from the Sun, 930,000 miles (1.5 million km) from Earth, which is around four times the distance to the Moon. This location makes it easier to do a whole sky survey without the obstacle of Earth in the way, and that reduces the interference usually picked up by a satellite from Earth-based radiation sources.

The Lagrangian points

In total there are five different locations where the gravitational pulls of the Earth and the Sun balance out, making it a stable location to host a satellite (not to scale).

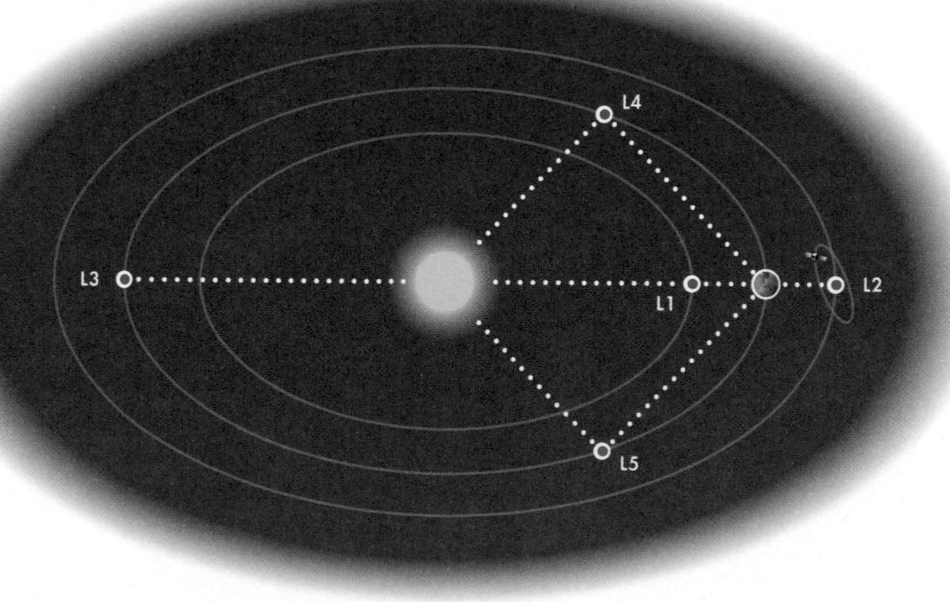

THE PATTERN OF REALITY

Looking at the oval images that WMAP and COBE produced, it's hard at first glance to see quite what got scientists so excited at the time. Stephen Hawking referred to COBE as "the greatest discovery of the century, if not of all time." This was pure hyperbole, considering that was a century when vast swathes of fundamental science, from quantum physics to relativity—and the Big Bang theory itself—was developed. However, there's far more to the data collected by WMAP, and its successor Planck, than meets the eye. That stretched egg-shaped image is just an overview of the data that WMAP collected over each six-month period that it took to scan the entire sky. This data collection, based on observing a thin strip of the sky at a time, is vastly more detailed.

WMAP's successor, the Planck observatory, was a European Space Agency satellite, launched in 2009. Positioned near L2 like WMAP, it made observations for four years, producing significantly more detailed scans of the cosmic microwave background than WMAP had achieved. As was also the case with WMAP, after finishing observations Planck was moved into a so-called graveyard orbit around the Sun, to vacate space in the valuable L2 location. It might seem wasteful to retire these satellites after a relatively short lifetime. The Hubble Space Telescope, by comparison, has been operating since 1990 (more accurately since 1993, when an initial fault was fixed) and may still be in use as late as 2040. However, the cosmic microwave detection satellites need ultra-cold devices—in the case of Planck it was necessary to sustain a temperature of 0.1 K (-459.49°F or -273.05°C), extremely close to absolute zero. This is only possible if a limited supply of coolant is available that evaporates over time.

Looking at the latest images from Planck, we can see what is effectively a pattern generated by the broad distribution of matter, practically all in the form of hydrogen atoms, in the early universe. In some parts of the early universe there would have been more matter atoms than in others. It might seem that there shouldn't be such clusters of matter in a randomly distributed collection of objects, but a random distribution is not an even one. It will have bunches and gaps. (If you doubt this, imagine dropping a box of ball bearings on the floor. It would look extremely suspicious if they were all evenly spread out in a regular grid.)

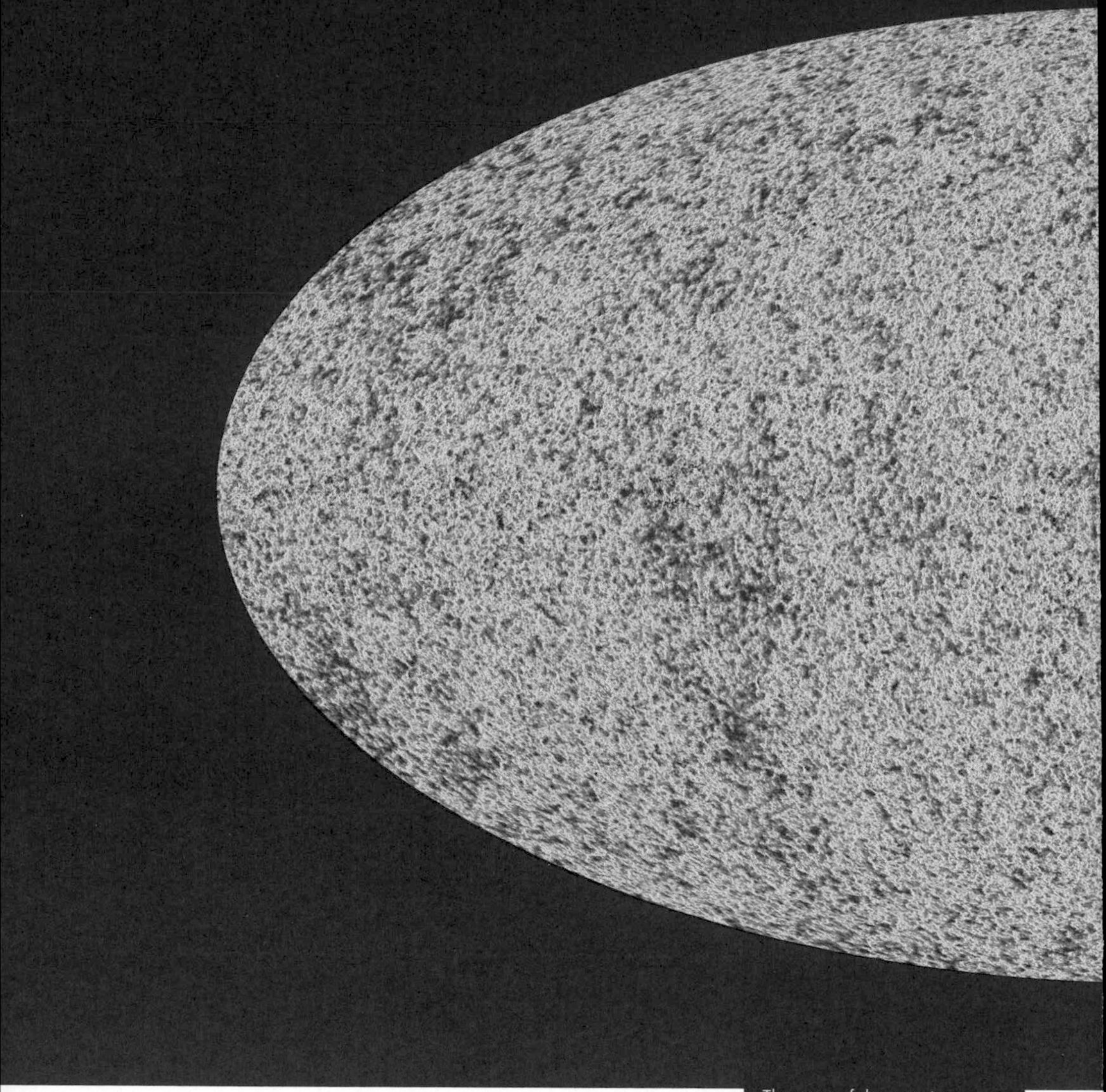

The map of the cosmic microwave background radiation from the Planck satellite shows the best detail yet recorded.

To add to this random nature, the universe started off extremely small, on a scale where quantum effects would have dominated. Quantum physics, which describes the behavior of matter and energy on a very small scale, is inherently probabilistic and would impose random variations on the distribution of energy and the matter derived from it that could have formed. These small and random variations, displayed in the cosmic microwave background, would form the seeds of the galaxies that over time would come together from scattered gas across the universe.

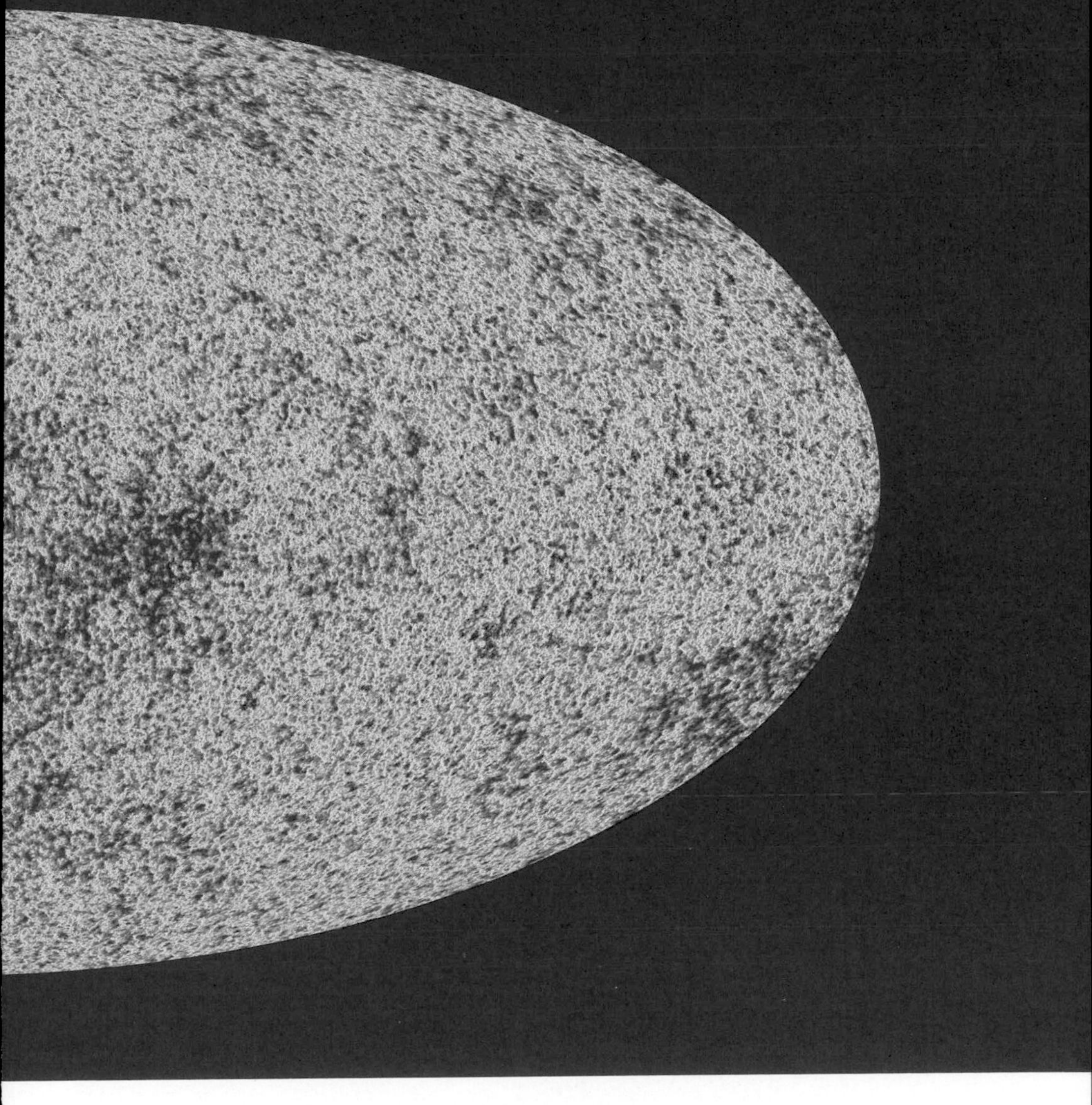

By studying the variation in this cosmic background radiation, a lot can be deduced about the structure of the early universe. For example, the dark energy that is driving the acceleration of the expansion of the universe could be consistent across time and space, or could vary. Observations from WMAP suggest it is more likely to have the consistent approach sometimes described as a "cosmological constant." Studying the cosmic microwave background radiation in detail also makes it possible to study the way that this radiation is affected by influences such as gravitational lensing—where massive bodies warp space and change the direction of electromagnetic radiation like a massive lens—

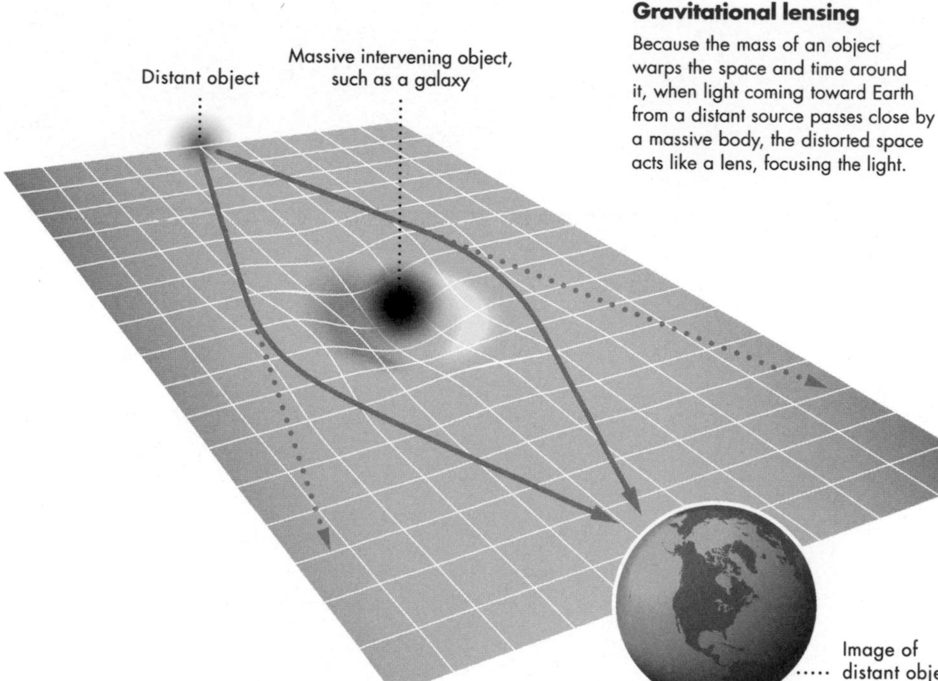

Gravitational lensing

Because the mass of an object warps the space and time around it, when light coming toward Earth from a distant source passes close by a massive body, the distorted space acts like a lens, focusing the light.

and has even provided information on bodies in the solar system, studying them indirectly through their influence as the radiation passes by.

Planck has gone beyond WMAP in providing more accurate data that, for example, can give more specific values for the ratio of matter to dark matter in the universe and for the age of the universe. Dark matter is a hypothetical substance that is thought to influence the way that galaxies spin around. It has mass, but does not interact with electromagnetic radiation, which means it can't be seen and usually passes through normal matter undetected. Although some astrophysicists think that dark matter doesn't exist, with the effects attributed to it caused by variations in the effect of gravity, most think that it is there, and if so, Planck suggests that there is around five times as much dark matter as ordinary matter in the universe. When analyzed, data from Planck also indicated that the universe was a little older than had previously been thought, pushing back the best estimate from 13.7 to 13.8 billion years.

LIMITATIONS AND FUTURES

The pattern in the cosmic microwave background can tell us much about the nature of the early universe, but it was hoped that a series of experiments called BICEP would also provide evidence of a hypothetical early phenomenon called inflation. This was devised to fix a problem with the Big Bang model, but as yet has had little confirmatory evidence to support it.

BICEP: This stands for Background Imaging of Cosmic Extragalactic Polarization.

As we have seen, the universe has expanded from the beginning, but the Big Bang theory, of itself, does not match what is actually observed. To patch up the theory, it was suggested that when the universe was less than one hundred millionth of a trillionth of a trillionth of one second old, it suddenly expanded to be at least one million trillion trillion trillion trillion trillion trillion trillon times larger than it had been. There is no real explanation for this "inflation" process, but it was necessary to make the theory match what was observed.

The BICEP experiments hoped to find evidence for inflation in the cosmic microwave background radiation. Light can be polarized: polarization is a measurable property of the light that indicates a particular direction at right angles to the light's direction of travel. If the light that made up the background radiation had a particular type of polarization, this would reflect an early movement of the energy in the universe that would be expected from inflation, helping confirm that inflation really happened.

In 2014, the second BICEP experiment, which has a dramatic location at the South Pole, made a discovery of a polarization pattern that seemed to strongly support the existence of inflation. However, a few months later the BICEP scientists were forced to retract their claim because it was likely that the effect was actually produced by the light passing through dust in space, which can also affect polarization. At the time of writing, inflation is still to be supported by evidence.

This in no sense takes away from the importance of the cosmic microwave background radiation. As we have seen, this has already given us much useful information about the early nature of the universe, and will continue to do so. However, the BICEP error emphasizes just how indirect any deductions from this

The Andromeda galaxy is the nearest large neighbor to our Milky Way galaxy. Located around 2.5 million light years away, the Andromeda galaxy contains around one trillion stars.

radiation are. We are looking at tiny variations in a low-level form of radiation that has been passing through space for almost the entire lifetime of the universe, interacting with and being distorted by all kinds of matter along the way. It is quite remarkable that this pattern has been able to tell us so much.

Whenever we look in to space, we look backward in time. Light, whether visible or at different energies, such as the cosmic microwave background, is the fastest thing in existence, but it does take time to get from A to B. This means that the greater distance we look out into the universe, the further back in time we are seeing. The most distant object visible in the night sky to the naked eye is the Andromeda galaxy, from which the light has been traveling for 2.5 million years. We see it as it was 2.5 million years ago. But the cosmic microwave background gives a view into the universe's far prehistory, looking 13.8 billion years back in time. This pattern remains a time tunnel to the past that will continue to tell us more about where our universe has come from.

By contrast, our next pattern gives us an insight into something very different. The fundamental and intertwined relationship between the two key elements of reality—space and time.

"TELESCOPES ARE TIME MACHINES. WHEN WE LOOK OUT AT SPACE, WE ARE LOOKING BACK IN TIME. NASA

2 MINKOWSKI DIAGRAMS

Hermann **Minkowski**
1864–1909

SPACE-TIME PATTERN

Albert Einstein transformed our understanding of reality with the special theory of relativity, showing that time and space are not independent, but part of a linked whole that would become known as space-time. Einstein's old math professor from Zurich Polytechnic in Switzerland—Hermann Minkowski—developed a new kind of diagram that shows how time and space are related and inseparable. These simple diagrams make the strange outcomes of relativity much clearer, showing, for example, how two apparently simultaneous events will not occur at the same time for moving observers, and forming the light cones that Stephen Hawking would use to demonstrate the effects of black holes. Roger Penrose would extend the diagrams to encompass an entire infinite universe.

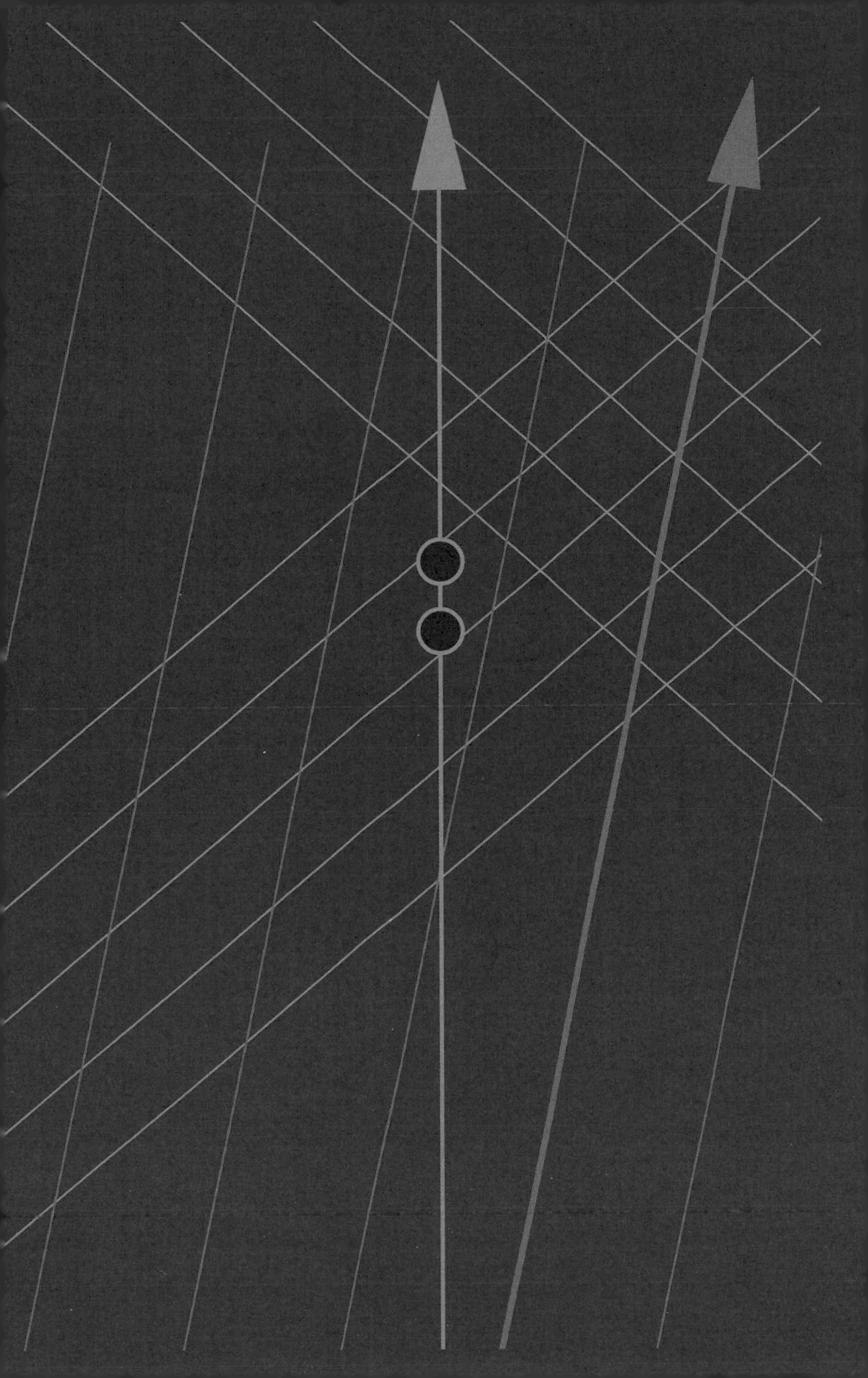

PATENT OFFICE DROPOUT

The story behind this space-time pattern begins in the Swiss Patent Office in Bern in 1905. The young Albert Einstein, then aged 26, had failed to get a university position, but found that checking patents was easy work that enabled him to spend a considerable amount of time on his own interests. In the breakthrough year of 1905, Einstein published the paper that won him the Nobel Prize—on the photoelectric effect that makes solar panels possible—which would also provide one of the foundations of quantum physics. But from the viewpoint of relativity, he wrote two other key papers that year. One provided his famous equation $E=mc^2$ (though not in that exact form). The other, to which the $E=mc^2$ paper was only an extension, established the special theory of relativity.

It is thought that Einstein was inspired to think about the relationship between space and time that lies at the heart of special relativity by some of the patents that were part of his day job. At the time, the advent of the railways meant that keeping the same time in different locations had become important. In the past, each town kept its own time, which could vary considerably from location to location. It wasn't possible to run a railway on such a temporally fragmented basis. As a result, Einstein had to check a number of patents for methods of electrically synchronizing clocks. One of

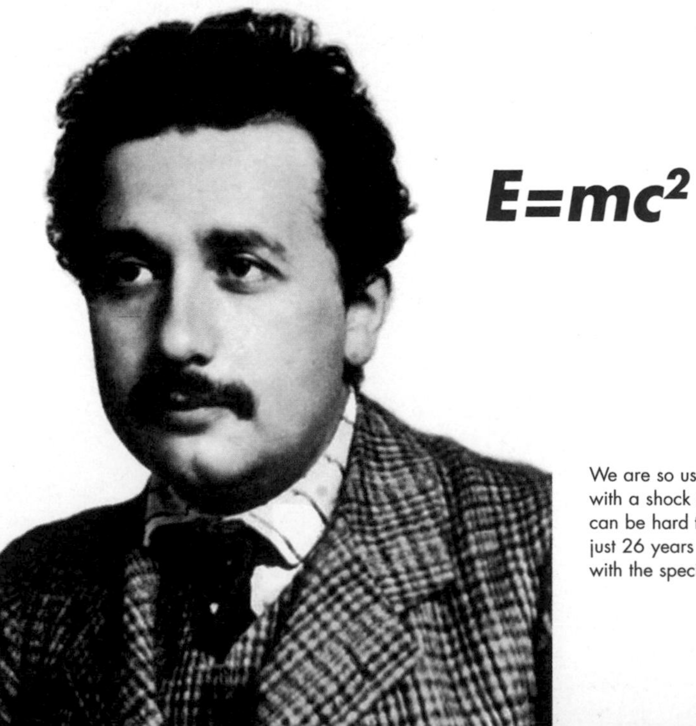

We are so used to seeing Einstein with a shock of white hair that it can be hard to remember he was just 26 years old when he came up with the special theory of relativity.

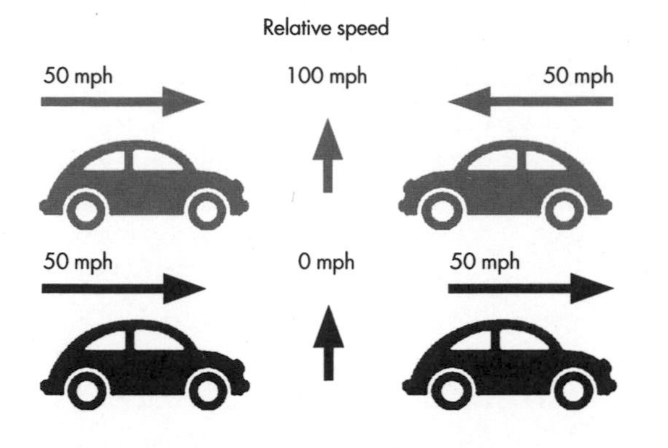

Relative motion

When two cars head toward each other at, say, 50 miles per hour (80 km/h), the relative speed is 100 miles per hour (161 km/h). When they head in the same direction, the relative speed is zero.

the earliest thoughts he had on special relativity was about the "relativity of simultaneity"—what it meant for two events, occurring at remote locations, to genuinely happen at the same time.

The traditional concept of simultaneity would be shattered by Einstein's paper on relativity, entitled *Zur Elektrodynamik bewegter Körper* (On the Electrodynamics of Moving Bodies). In it, Einstein pulled together ideas that had been discussed for the few years leading up to 1905 to make a remarkable transformation of our understanding of space and time. This was a result of combining two apparently unrelated observations. The first was Galilean relativity, established in the seventeenth century and showing how moving objects behaved.

Providing the foundation for Newton's laws of motion, Galileo had shown that all motion is a relative concept. We can't say that something is traveling at 60 miles per hour (96.5 km/h) without establishing with respect to what it is moving. Because Earth dominates our view, we tend to measure speed with respect to the Earth's surface, but that is a purely arbitrary "frame of reference" as physicists call it.

Apart from anything else, the Earth is flying around the Sun at 18.6 miles per second (30 km/s), so it is not exactly a stationary starting point. Similarly, if we imagine two cars are driving along a road, each traveling at 50 miles per hour (80 km/h). If the cars are heading toward each other, relativity tells us that their combined speed toward each other is 100 miles per hour (160 km/h). If they are both heading in the same direction their relative speed is zero.

THE NATURE OF LIGHT

The other input to special relativity came from the work of one of Einstein's heroes, the Victorian Scottish physicist, James Clerk Maxwell. It was Maxwell who had discovered the nature of light—an interplay between electricity and magnetism. Light's formation from the interaction of changing electrical and magnetic fields could only happen if light had a specific speed for the medium it travels through. We now know that speed to be exactly 299,792,458 meters per second (around 186,000 miles per second) in a vacuum. (We can know it precisely because the meter is defined as 1/299792458 of the distance light travels through a vacuum in a second.)

This fixed speed of light cannot be influenced by relativity. If it was, every time we moved with respect to a beam of light, the light would disappear, as it would no longer be traveling at the only speed at which it can exist. So, unlike anything else, moving toward or away from a beam of light has no effect on the speed at which the light travels toward you.

When this oddity of light is plugged into the equations of motion based on Galilean relativity, the result is several surprising outcomes. It shows that when an object is moving with respect to you, you will see the time slow down on the object; it will shrink in the direction that it is moving; and it will increase in mass.

This doesn't seem an obvious implication of relative motion, but we can see without math how the time effect works by considering a device known as a light clock. This is a clock where the equivalent of a pendulum in a conventional clock is a beam of light dashing up and down between a pair of mirrors. Each "tick" of this clock is when the light hits one of the mirrors.

Imagine we put such a clock on a spaceship (see opposite) and sent that ship off at high speed into space. To someone on the ship, the clock would be unaffected by the movement. The beam of light would continue to bounce up and down in a straight line. This was even predicted by Galileo, who showed that inside a steadily moving sealed vessel (without looking outside) there was no possible experiment that would prove that the vessel was moving.

Now let's look at the spaceship from Earth, starting from the point in time when the beam of light is leaving, say, the top mirror. As the ship moves forward, the beam would not travel vertically down, but would travel at an angle so that it hit the bottom mirror, by now moved with the ship, at the end of the light beam's travel. The light would have to cover a greater distance than it does for someone on the ship. As light's speed can't alter, the only way this would be possible is if time was running slower on the ship, giving the light the chance to get to the second mirror.

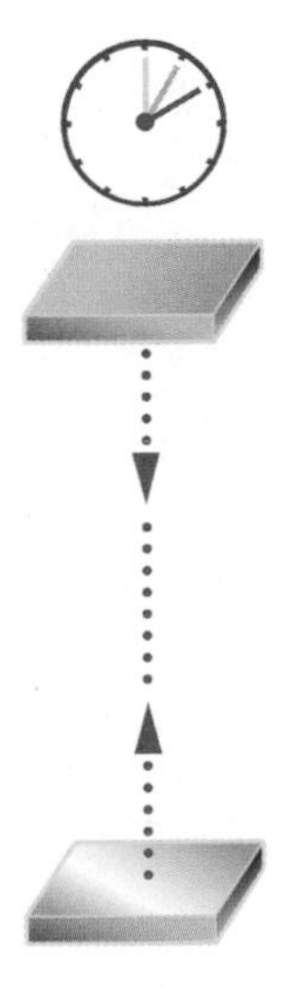

No movement

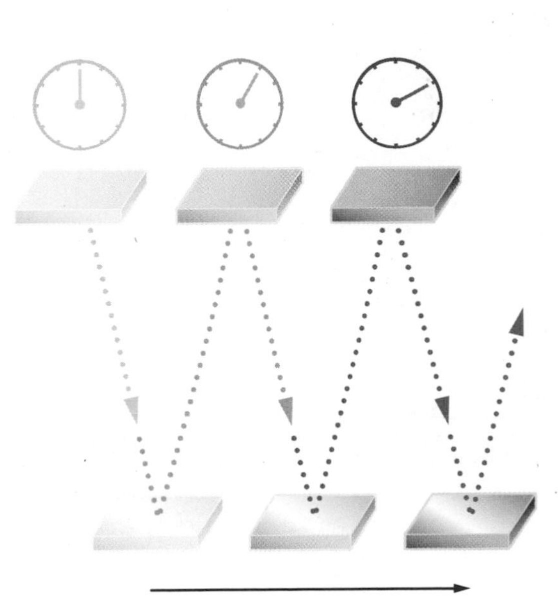

Direction of travel

Light clock on a spaceship

On the spaceship, light bounces straight up and down, traveling the distance between the mirrors. But seen from Earth, the light travels further, moving diagonally as the mirrors move with the ship.

The Crab Nebula is the remnant of a supernova that occurred around 5400 BCE, but was not seen from Earth until 1054 CE.

FROM THEORY TO DIAGRAM

What the special theory shows is that time and space are not two separate entities. Movement in space influences time. As far as Einstein was concerned, the mathematics were all that was needed. He was a pure theoretician (his only venture into practicality was co-designing a new type of fridge). But his old university math lecturer, Hermann Minkowski realized that it would be useful to have a visual representation of the relationship between time and space. Einstein was initially not at all convinced about this, but eventually came to recognize the value of this pattern, the Minkowski diagram.

EINSTEIN'S FRIDGE: Working with Leo Szilard, Einstein designed a fridge with no moving parts that needed only a heat source to work.

At its simplest, a Minkowski diagram is a chart where the vertical axis is time and the horizontal axis is space. (There are, of course, three dimensions of real space, but for simplicity we can just consider motion in one arbitrary direction.) On the diagram we can plot the position of anything in space and time. For example, a stationary object would be shown as a vertical line heading up in the time direction from past to future.

By contrast, a steadily moving object would travel along a diagonal path, with the slope of the line dependent on the speed at which the object moved. The plot of the position of an object in space and time on a Minkowski diagram is known as its world line.

Most real things—like you, for instance—do not move around at a constant speed, so your world line would shift around the diagram, always heading upward, but sometimes moving straight upward as you stood still, at other times following a curve as you moved around.

For convenience, the distance along the horizontal axis is best measured in light time units. The most familiar of these is the light year. Despite its misuse as a unit of time in numerous science fiction films (notably, for example, in *Star Wars*), one light year is a unit of distance. It is the distance that light travels in one year. As we've already seen, light zaps along at 299,792,458 meters per second. One year is around 31.5 million seconds, making one light year the equivalent of around 5.88 trillion miles (9.46 trillion km).

We can equally measure distance in light days, light hours, or light seconds. The Sun, for example, is around 499 light seconds from Earth as it takes our neighborhood star's light 499 seconds to reach us (so we see the Sun as it was 499 seconds ago). When drawing a Minkowski diagram, we use the same light time measure of distance on the horizontal axis as we do units of time on a vertical axis. So, if we're measuring time in seconds, we measure distance in light seconds; if we're measuring time in years, we measure distance in light years.

The result of using these units is that a beam of light travels at 45 degrees up a Minkowski diagram, because, by definition light travels (say) one light second in one second. If we add two lines, traveling at light speed in opposite directions, this divides the diagram into an important pattern of space and time.

Static world line

A static object's location in space-time travels through time, but not through space. As a result, its world line is parallel to the vertical axis.

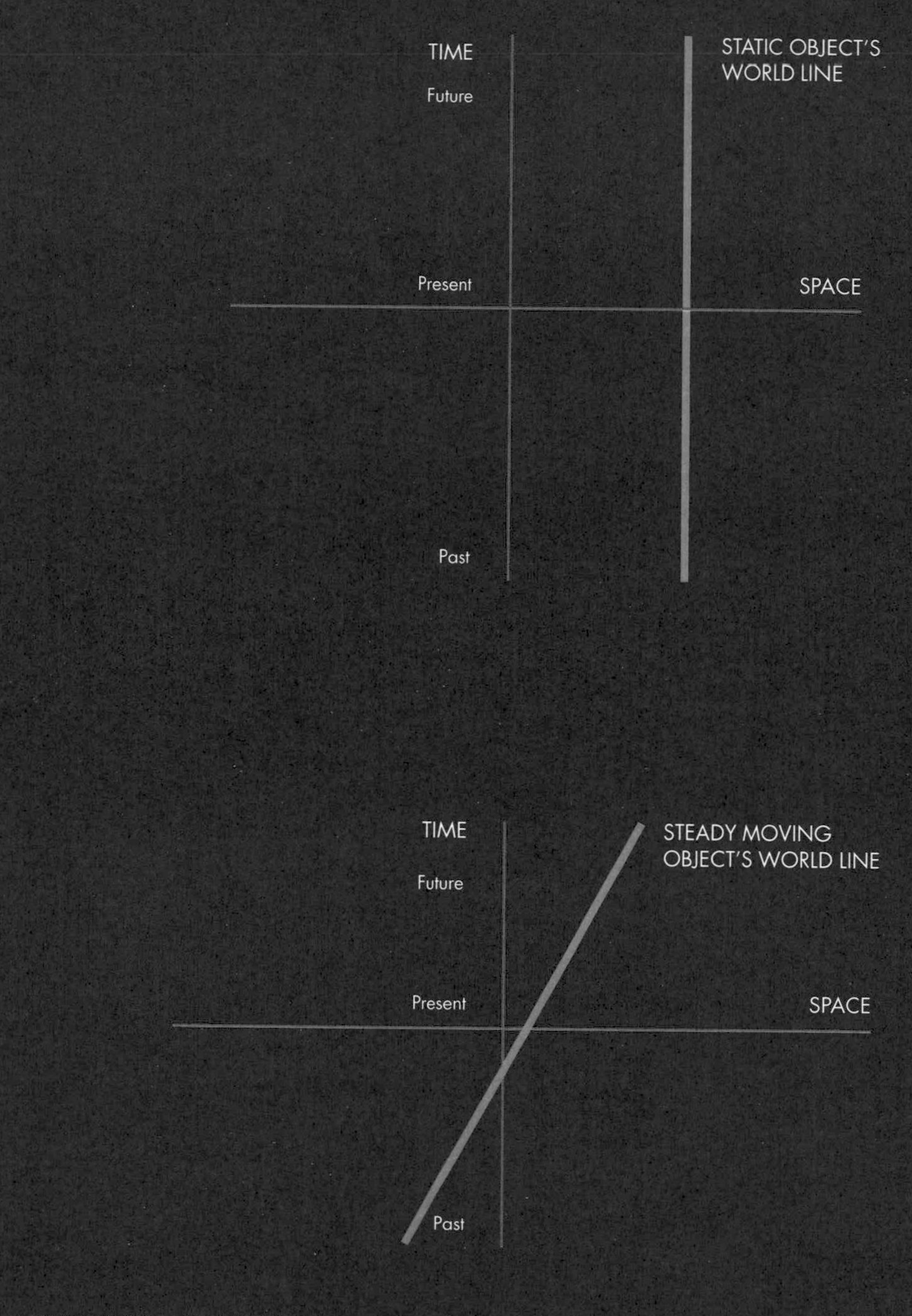

Moving world lines

A moving object's location in space-time travels through both time and space. At a steady speed this produces a straight world line, while acceleration makes the line curve. If distance is measured in light years and time in years, a light beam travels at 45 degrees to the horizontal and vertical axes.

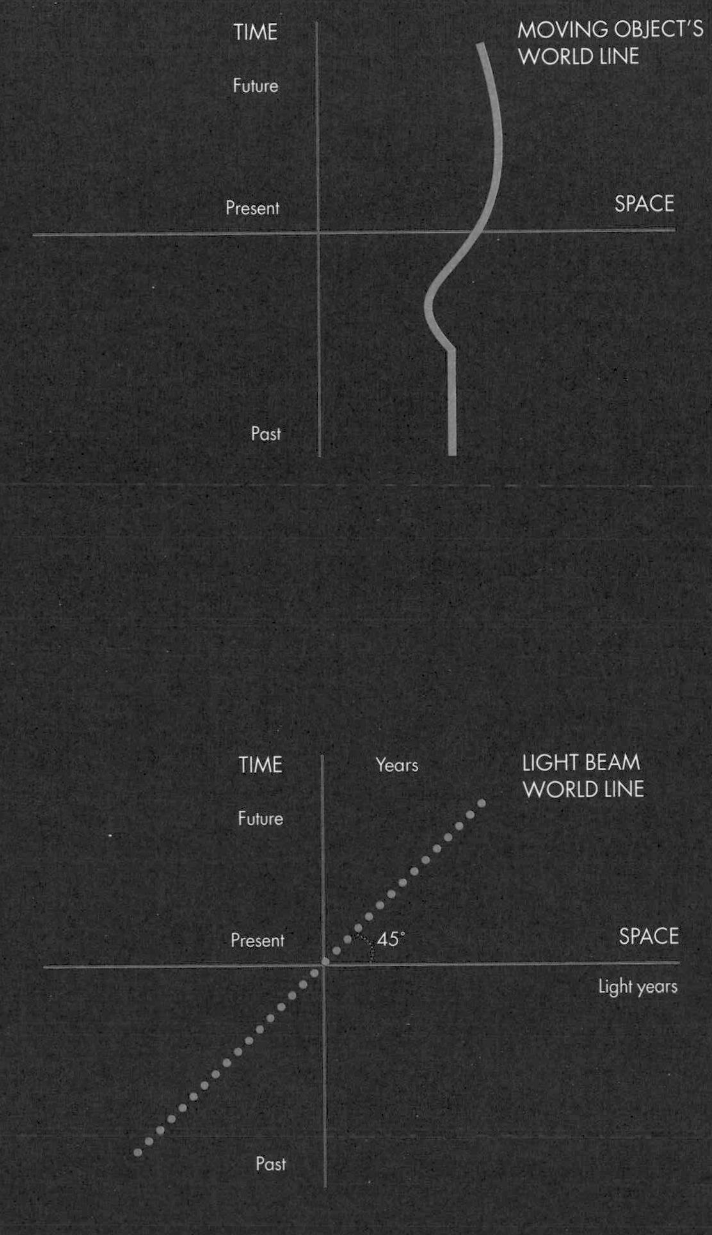

REALITY AND THE ELSEWHERE

The interior of the blue cross part of the diagram (see opposite top) is effectively the real world, the world in which it is possible to exist. The part of the diagram outside the blue cross is a forbidden zone. For a world line to pass through this zone it must be moving through space faster than light does. But this is impossible. Remember that the special theory shows that when an object moves, time slows and its mass increases. As an object nears the speed of light, time comes toward a standstill and its mass gets closer and closer to infinite. This means that it would take an infinite amount of energy to accelerate it past the limit. As this can't happen, nothing can enter the forbidden zone.

The region inside the upper part of the blue cross is the future and can be influenced by whatever happens at the pinch-point of the cross—the present. The region inside the blue cross's lower half is the past and can't be influenced by an event in the present (though events in this region can influence the present and future). And the region outside the blue cross would require communication faster than the speed of light for any influence to travel from the present to them, or from them to the present, so they are isolated and sometimes referred to as neither the past nor the future but the "elsewhere."

If we look at a slightly more complex Minkowski diagram (see opposite bottom), we can see how an event can move from the elsewhere into the real future. Imagine that a vast supernova explosion took place ten light years from Earth. On the diagram, we are looking from the Earth's viewpoint. For the first ten years, the explosion is in the elsewhere. It can have no effect on Earth. But after ten years, the event moves into Earth's real future. Initially light, and soon after matter traveling at close to the speed of light, from the supernova could arrive at Earth.

The simple pattern of the Minkowski diagram also allows us to get a better understanding of the new take on the concept of simultaneity that was forced on us by the special theory of relativity. As we have seen, one of Einstein's earliest ideas in this field was trying to work out what it meant to say that two events at remote locations took place simultaneously.

Plotting the elsewhere

The region outside the 45-degree lines of light speed is a forbidden zone where there can be no influence.

When an event, such as a supernova, occurs in the elsewhere (see bottom diagram), it can't be detected until its world line takes it above the light line.

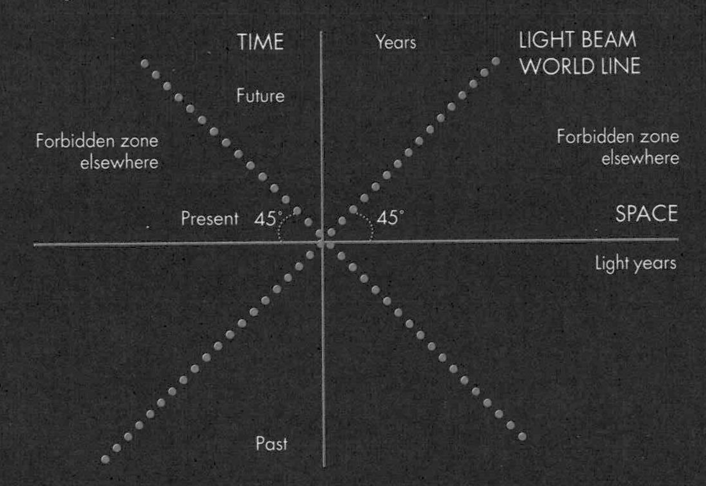

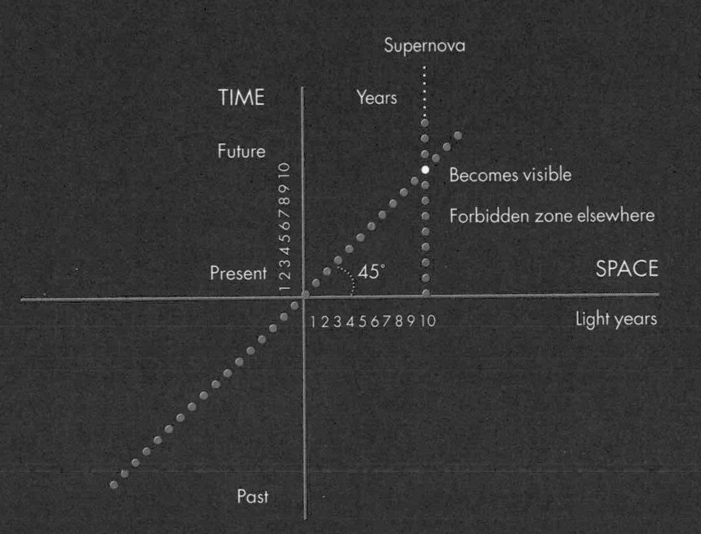

THE RELATIVITY OF SIMULTANEITY

Einstein described an imaginary situation where two lightning strikes simultaneously hit a straight railway line at two different locations. (This is a so-called thought experiment, one of Einstein's most frequently used tools for exploring reality.) What does it mean, to say that these events are simultaneous? How can we even know that this is the case, as we can't be in both places at once to observe the lightning?

Einstein imagined an observer—let's call her Lucia—conveniently placed exactly midway between the two locations. (This is why what we're dealing with can only ever be a thought experiment, it could never be done in reality because in the unlikely event of simultaneous lightning strikes hitting the same straight railway line, we wouldn't know where they were going to be in advance, so we couldn't put Lucia in the middle.) Placed at the exact midpoint, Lucia could look in both directions (or, more realistically, she could have detectors pointing in both directions) and note the time at which the light from the strikes arrived at her location.

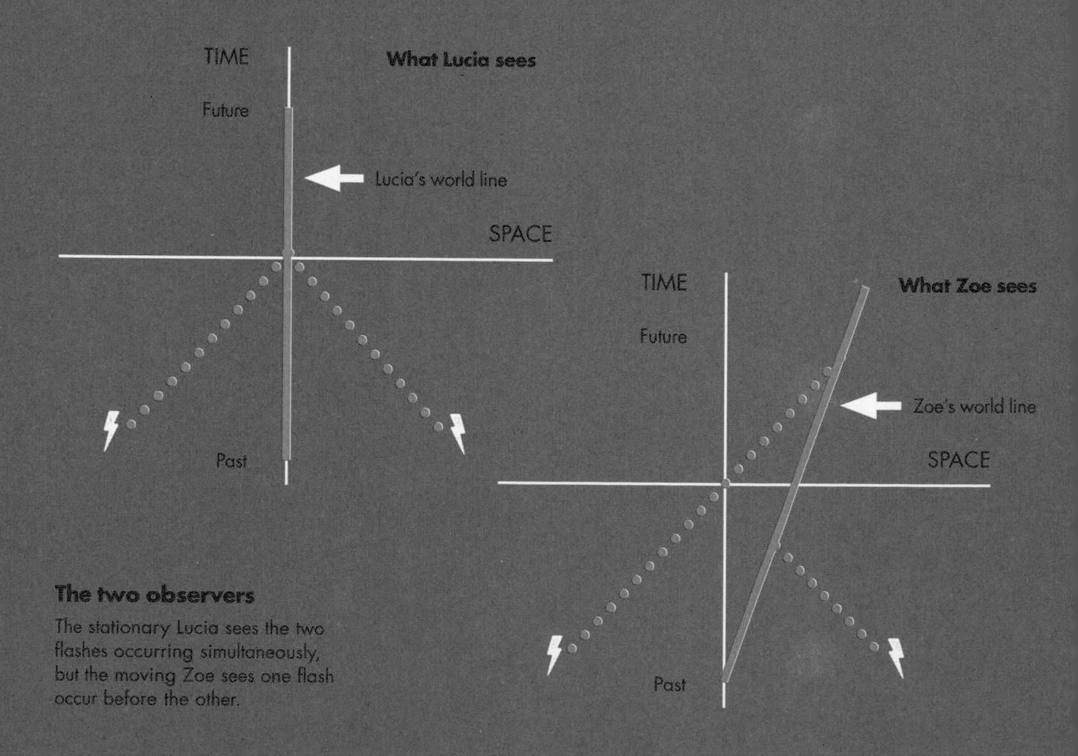

The two observers

The stationary Lucia sees the two flashes occurring simultaneously, but the moving Zoe sees one flash occur before the other.

There is one assumption here. That there weren't any significant obstacles in the way—light travels slower in more dense substances; so, should the light have to travel through glass in one direction, but not in the other, it would arrive a little later after passing through the glass despite the events being simultaneous. But given this assumption, Lucia would say from her viewpoint that the events were simultaneous if the light from both the lightning strikes arrived at her midpoint location at the same time.

However, let's look at a small variant on the experiment, which is why it was located on a (straight) rail track in the first place. Imagine that a second observer, Zoe, with the same equipment, is located on a high-speed train, passing along the tracks. As even more of a coincidence, Zoe passes Lucia at exactly the time when Lucia observes the simultaneous events from the trackside. What would Zoe see from her position on the train?

Imagine the train is running from left to right. The speed of the train does not add to or subtract from the speed of light coming from either lightning strike—that's the way light works. But in the time that it takes light to get to the train, Zoe and her detectors will have moved. This means that the light from the right-hand strike will arrive at Zoe's detectors *before* the light from the left-hand strike. For Zoe, the events are not simultaneous.

It can be difficult to understand exactly how this all works in practice—which is where the pattern of the Minkowski diagram comes in. We can look at what happens from the viewpoint of the two observers.

The diagram (opposite) makes it much clearer to comprehend how Lucia and Zoe will see the same events in a different way. The pattern of simultaneity is shattered by motion, one of the key results of the special theory of relativity.

The Minkowski diagram was expanded in the 1960s by two physicists, the Englishman Roger Penrose and the Australian Brandon Carter. The modification they made to the original form involves a type of projection. But before getting into the specifics of the Penrose (or Penrose-Carter) diagram, we first need to get a picture of what mathematical projections involve.

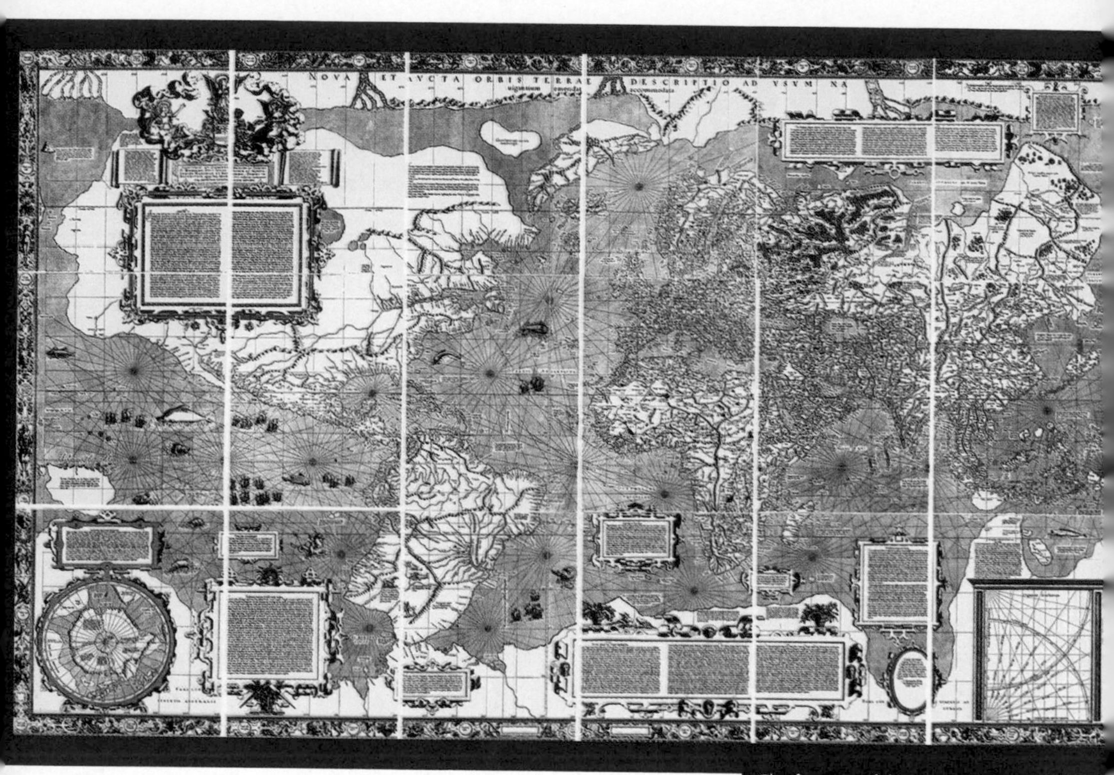

The first use of the Mercator projection by the Flemish mapmaker Gerardus Mercator was in this world map created circa 1569.

PROJECTING INTO FLATNESS

The most familiar projection is that used to draw the flat map of the world. The Earth is (roughly) spherical, so does not lend itself to easily being drawn flat. The approach taken is to imagine each point on the Earth's surface being mapped onto a flat sheet according to some (relatively) simple rules. This mapping process is known as projection. The best-known map projection of the Earth is the Mercator projection, which was first used by the Flemish mapmaker Gerardus Mercator in 1569.

PROJECTION: We tend to think of projection as showing a large image, but mathematically it is the result of drawing lines from a fixed point through the source to the destination.

To make a Mercator projection, we imagine a cylinder with the same diameter as the Earth (as shown overleaf). The planet is positioned in the middle of the cylinder. One way to envisage the projection process is that there is a light inside a transparent version of the Earth, and we project (in the sense of a slide projector) the points of the Earth's surface onto the cylinder. In the particular way this is done for Mercator, meridians (lines running from the North to South Pole) taken at equal angles become equally spaced vertical lines on the map, while the lines of latitude become horizontal lines on the map. The result is the familiar layout—it is quite an accurate representation of reality near to the equator, but becomes increasingly inaccurate as we move toward the poles.

Although variants of the Mercator projection are the most frequently used, there are many possible alternatives. For example, it is possible to have a projection that forces the areas of different regions on the map to be in the correction proportions, although "equal area" maps distort the shapes of countries and continents.

Alternatively, there are forms of projection where the distances from a selected point on the map to anywhere on the globe are accurate. A common approach here is the azimuthal equidistant projection where the map is centered on a selected point, increasing in distance from that point to the edge of a circular map. In such a projection, the entire edge of the circle is effectively the point on the opposite side of the globe from the selected point, which can make it look quite disconcerting. So, for example, if the map is centered on the North Pole the rim of the circle is the South Pole.

"THE DISTINCTION BETWEEN PAST, PRESENT, AND FUTURE IS ONLY A STUBBORNLY PERSISTENT ILLUSION.

ALBERT EINSTEIN

Map projection

When making a map like Mercator's, the Earth's features are projected onto an imaginary cylinder, as if there were a light at the center of a transparent planet.

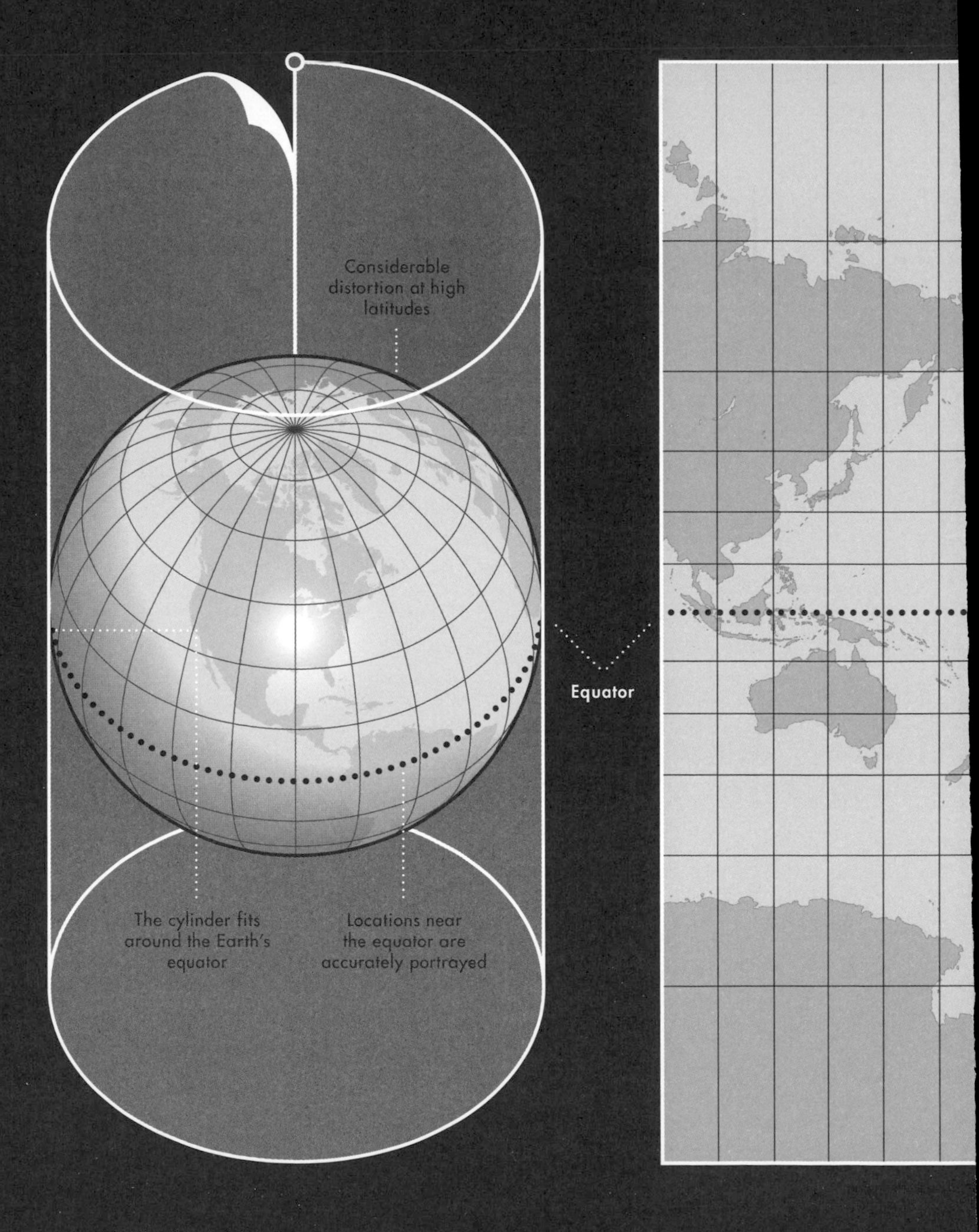

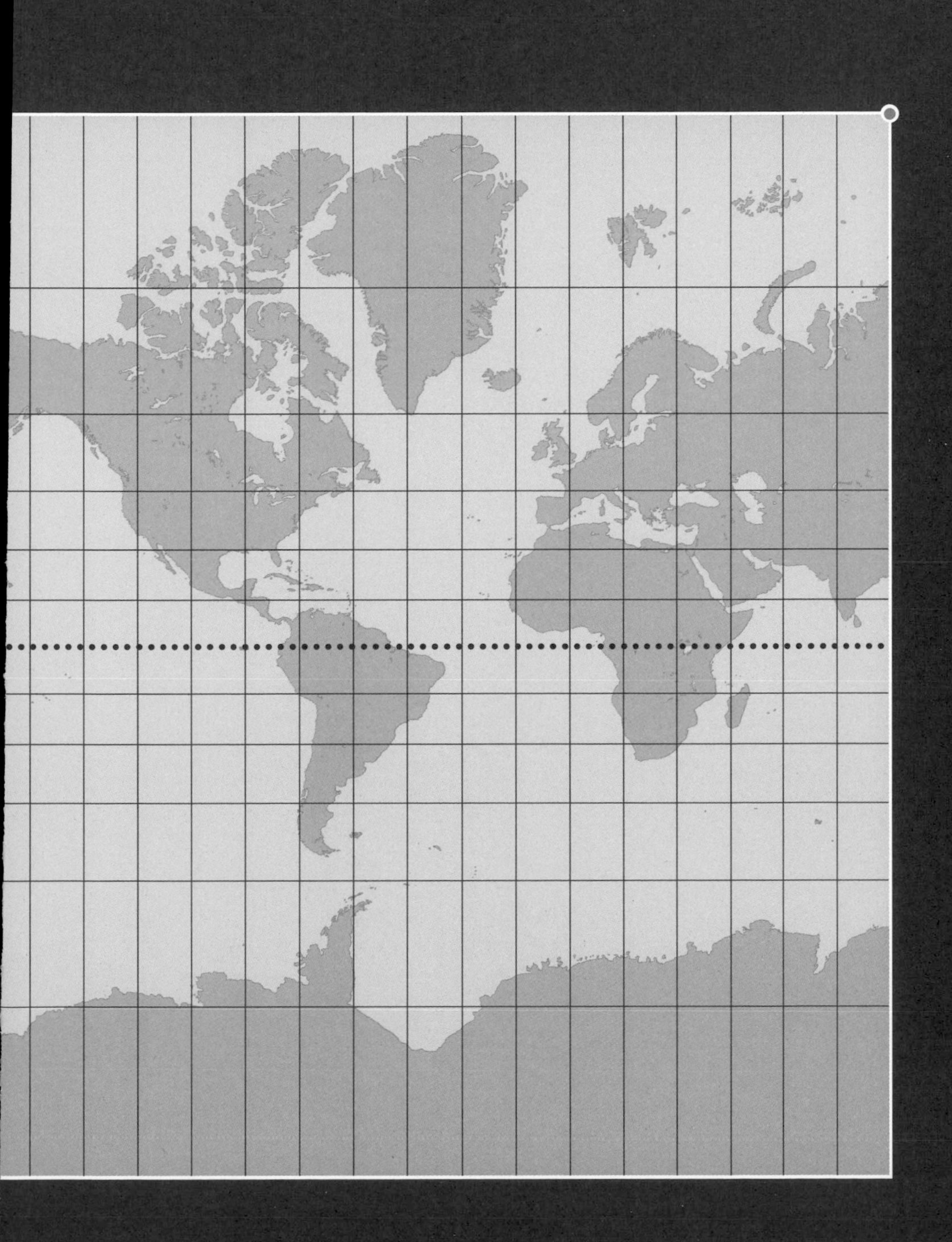

SCRUNCHING UP INFINITY

In a Penrose diagram, a similar kind of approach is taken in order to be able to map an entire infinite universe onto a finite diagram. To do this, the further an object is from the current time and location, the more the distance on the diagram gets scrunched up. The rate at which the degree of "scrunched-upness" increases is fast enough to be able to represent an infinite space and time on a finite diagram.

ENVISAGING INFINITY: We can struggle with the concept of infinity, but mathematically it is a powerful tool as we can often take an infinity collection of values to a finite limit, the process behind the mathematics of calculus.

If this seems impossible, imagine that for each unit of measurement on one axis of the Penrose diagram we half the distance on the diagram needed to show it. So, for example, we might use 1 inch on the diagram to represent the first light year away from Earth into space. Then the next half inch would represent the next light year, followed by a quarter inch for the next light year—and so on, forever.

Penrose diagram

A form of Minkowski diagram where each of the axes gets exponentially smaller, so all possible values are contained in a small amount of space.

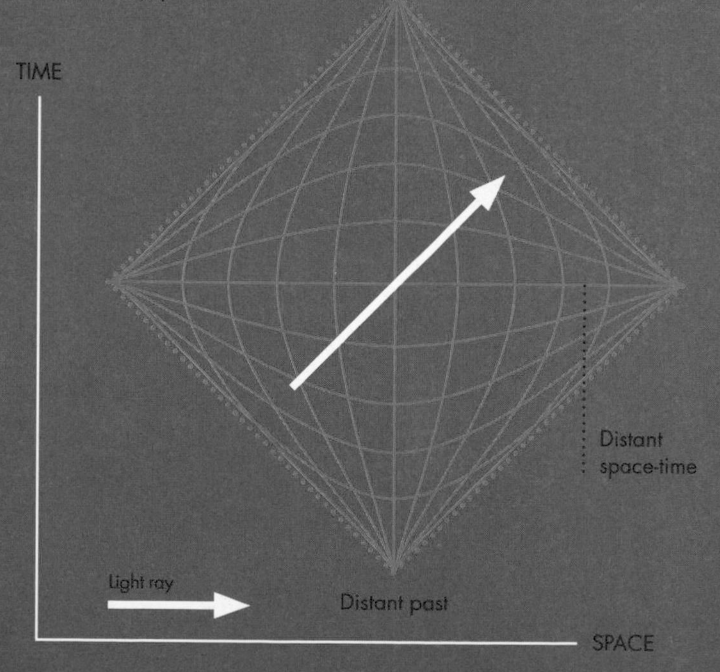

EXAMPLE OF AN INFINITE SERIES

An infinite series can be represented by:

$1 + \frac{1}{2} + \frac{1}{4}$... and so on.

The first three items in the series total $1\frac{3}{4}$.

We then get:

$1 + \frac{1}{2} + \frac{1}{4} + \frac{1}{8}$ totaling $1\frac{7}{8}$.

And so it goes on.

With n items in our series, the total is $1 + (2^{n-1}-1)/2^{n-1}$.

As n gets bigger and bigger, that total gets nearer and nearer to 2, effectively reaching 2 when we are projecting an infinite distance. So, using this projection, the entire infinity of time and space fits into a shape just 4 inches across (2 inches in each direction). Because of the specific mapping structure used in a Penrose diagram, the outcome is a diamond-shaped pattern.

Such diagrams proved valuable in plotting out the strange geometry of black holes, where what appears to be a simple sphere has such a massive distortion of space-time within it that it effectively stretches to infinity.

The Minkowski diagram and its derivatives show us the patterns of the relationship of space and time and, in the case of the Penrose diagram compresses something unimaginably huge onto a finite space. However, our next pattern concerns the interactions of the very smallest things in existence—the particles than make up reality.

“BLACK HOLES MAY BE APERTURES TO ELSEWHEN.

CARL SAGAN

3
PARTICLE TRAIL PATTERNS

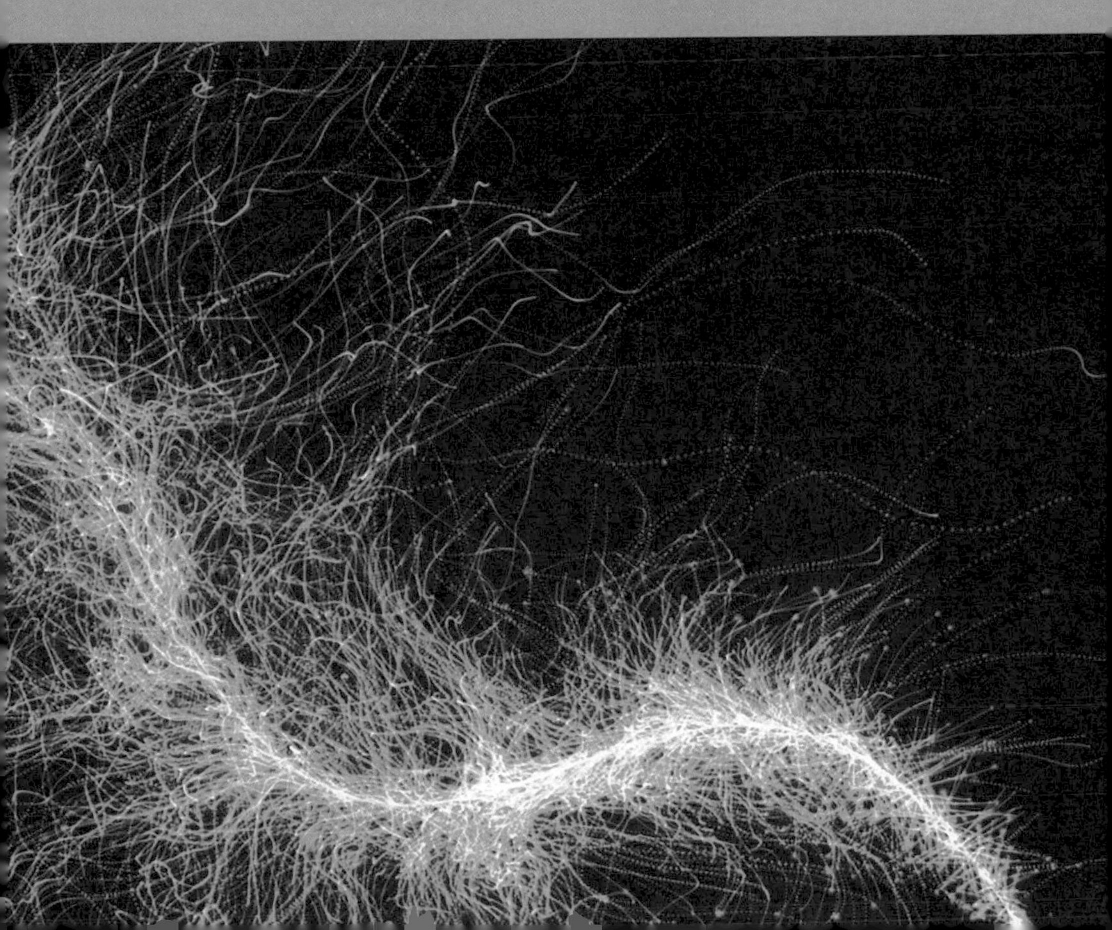

Peter **Higgs**
1929–

PARTICLES

Our understanding of the particles that make up everything around us has been built using accelerators and imaging chambers: devices that smash particles into each other at close to the speed of light and reveal passing particles, allowing us to study them. The most impressive accelerator to date, the Large Hadron Collider at CERN in Switzerland, is the biggest machine in the world. When a collision occurs in an accelerator, a complex mess of particles is produced, recorded in exotic-looking images. The patterns in these images have allowed scientists to do everything from discover the existence of antimatter to finding the elusive Higgs boson, named after Peter Higgs who theorized its existence. These are the patterns that uncover the fundamental building blocks of nature.

ATOMS, ELEMENTS, AND PARTICLES

What stuff is made from at the most fundamental level has fascinated humans for well over 2,000 years. The Ancient Greeks had two opposing ideas. The atomists believed that there were separate building blocks that everything else was made from. They imagined taking an object—for example, a block of cheese—and cutting it up into smaller and smaller pieces. Eventually, they thought, those pieces would be so small that they would be uncuttable, or "atomos"—atoms.

The standard model

Between them, these 17 particles are responsible for matter (quarks and leptons) and forces (gauge bosons), with the addition of the Higgs boson, from the field that gives some particles their mass.

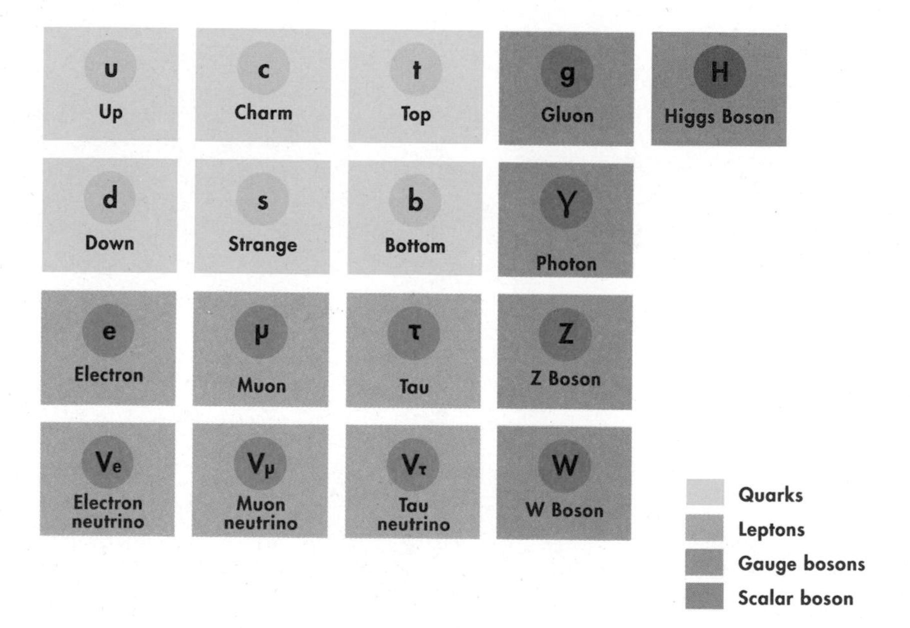

By contrast, other philosophers believed that everything was made from the elements of earth, water, air, and fire, and that these elements were continuous. Aristotle, for example, championed the four elements (with a fifth, the quintessence, for the heavens), as he believed that atoms required a void between them, which his view of physics did not permit. For a long time, Aristotle's viewpoint won out. However, in the nineteenth century the atomic theory was revived, first as a way to understand chemical reactions and, by the start of the twentieth century, as a representation of actual objects that made things. Objects, it was felt, had a structure of atoms.

If the name "atom" were literal—uncuttable—then that would be the end of it. But a series of discoveries from the end of the nineteenth century through the middle of the twentieth uncovered subatomic particles—unbelievably tiny specks of matter of which everything else consisted. Modern physics gives us a picture of a reality that is dominated by fields or particles.

Fields and particles are effectively inverse ways of looking at the same thing. A particle is a tiny concentrated speck of something, whether it be matter or energy. A field is like a contour map of something, filling all space and time. Sometimes there are tiny, localized peaks in the value of a field that can travel around. Such peaks are the things we call particles.

Our current "standard model" of particle physics gives us a picture of 17 different particles, although in practice we only tend to directly encounter four types in ordinary matter, plus the impact of two others in the forces of nature. This model itself makes for a mostly neat and well-ordered pattern.

It's entirely possible to dream up a model of how particles behave based on theory without any connection to reality. For example, in the early days of atomic theory, some thought that the atoms of different elements had various shaped hooks sticking out of them that enabled them to join together to form chemical compounds. This was an indirect guess and proved incorrect. But how can we know about these particles, which are far too small to see? The answer is primarily the result of the effect of the patterns of trails that particles leave behind them.

SEEING INTO THE NUCLEUS

One of the very first attempts to probe subatomic structures came when scientists working with Ernest Rutherford at Manchester University in England, discovered the atomic nucleus. It was already known that the atom contained smaller particles called electrons, which carried an electrical charge, but not how the whole thing fitted together. Rutherford's colleagues, Hans Geiger and Ernest Marsden, set up a series of experiments where particles called alpha particles (now known to be the nuclei of helium atoms) were fired at a thin piece of gold foil.

END OF THE "PLUM PUDDING": Before Rutherford's experiment, it had been assumed that electrons were scattered through an insubstantial, positively charged medium, the so-called "plum pudding" model where electrons were the equivalent of the fruit in a Christmas pudding.

The expectation was that some alpha particles would be deflected a little by the positively charged medium that was expected to contain the electrons, but in practice most traveled straight through, while a few bounced back. Rutherford realized that this was because the atom contained a compact, positively charged nucleus, dense enough for the alpha particles to bounce off. The term "nucleus" is now probably most associated with the atom, but Rutherford borrowed it from biology, where it was already used to denote a compact central structure in some types of cell.

Alpha particles themselves are far too small to see. As a result, Geiger and Marsden used a fluorescent screen—a sheet of material covered in a substance that gave off a tiny flash of light when it was hit by an alpha particle. This screen was wrapped around the gold foil in the dark and the researchers looked for a pattern of flashes to see where the alpha particles ended up.

Using a fluorescent screen gave an idea of where those particles arrived, but it was frustratingly limiting. It gave no clue of how particles got from A to B, merely denoting their end point. This is more than just a matter of determining a particle's route. Getting a picture of the journeys that particles take is essential in finding out more about them. It has even helped demonstrate fundamental aspects of science from the special theory of relativity to the mechanism by which most particles get their mass.

Gold foil experiment

In Rutherford's experiment, particles were observed to be deflected at a large angle by nuclei of gold atoms.

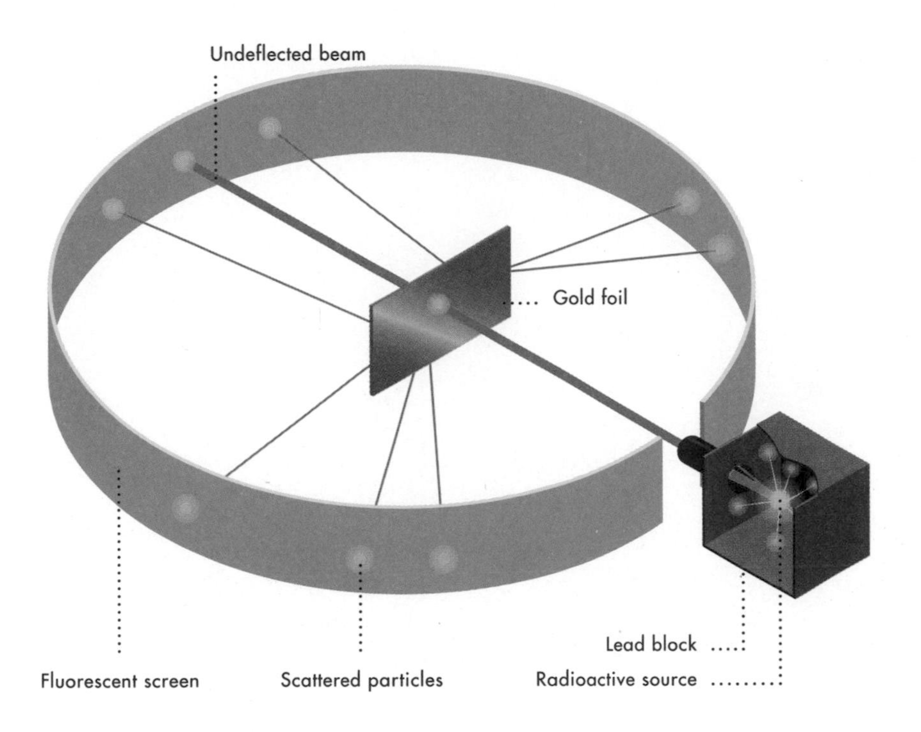

"IT WAS ALMOST AS INCREDIBLE AS IF YOU FIRED A 15-INCH SHELL AT A PIECE OF TISSUE PAPER AND IT CAME BACK AND HIT YOU.

ERNEST RUTHERFORD

In a cosmic ray shower, high-energy particles from deep space hit our atmosphere, triggering a cascade of new particles to be created.

TIME-TRAVELING PARTICLES

The special theory of relativity, which we met in the previous chapter, provided an impressive example of how knowing more about the pattern of a particle's progress gave an insight into a fundamental physical concept. The planet Earth is constantly bombarded with streams of high-energy particles from outer space, known as cosmic rays. (The term "rays" was often applied to particles in the early days of modern physics, and in some cases it has stuck.) When these particles hit the upper atmosphere they collide with atoms at speeds near that of light, resulting in the production of a cascade of new particles known as an air shower.

These particles were not present before the cosmic ray arrived. What has happened is a dramatic demonstration of one of the implications of special relativity, the world's best-known equation $E=mc^2$. This shows the equivalence of energy (E) with mass of matter (m). Here, the c refers to the speed of light—a big number as it takes a lot of energy to make matter. The cosmic ray particles collide with so much energy that a whole range of new matter particles are generated.

Among these newly formed particles are muons. These are, in effect, overweight electrons. In the standard model diagram on page 56, muons are in the same family as electrons, but are more than 200 times heavier. We don't encounter muons in the same way as we do the electrons that are responsible for both electrical currents and chemical reactions because muons are unstable. They have a typical lifetime of around two microseconds, after which they decay. This normally involves the muon changing to an electron plus two neutrino particles.

It was through observing the paths of muons from cosmic rays that impressive evidence of the special theory of relativity was gleaned. The muon paths were too long. As we have seen, relativity predicts that a moving object's time slows down. The muons were traveling sufficiently close to the speed of light that their time was slowed down significantly, enabling them to travel a lot further than they should have been able to before they decayed.

Yet how was it possible to discover this? Muons are invisible. They are far too small to see using any equipment. Yet the discoverers of the muon back in 1936 were able to reveal this effect of special relativity only a year later. And the discovery itself needed researchers to be able to deduce information about the mass and electrical charge of a particle they couldn't see. This became possible due to a clever way of producing a new kind of pattern, called a cloud chamber. This device for making the invisible visible was inspired by a ghostly phenomenon linked to a German mountain with a legendary past.

> “
>
> **THE MASS OF A BODY IS A MEASURE OF ITS ENERGY CONTENT.** ALBERT EINSTEIN

CLOUDS AND THE BROCKEN SPECTER

The mountain in question is the Brocken (sometimes called the Blocksberg), part of the Harz mountain range in northern Germany. At just 3,743 feet (1,141 meters) high it is not a dramatic summit, and is dwarfed by the larger European mountain ranges. However, despite its relatively diminutive size, the mountain has a legendary past. Like Pendle Hill in England, the mountain was traditionally linked to witchcraft, mentioned in this context by Goethe in his version of the Faust legend. Two rock formations on the mountain are popularly known as the Devil's Pulpit and the Witches' Altar.

Whatever the realities of the Brocken's dark past, it has given its name to a phenomenon known as a Brocken specter. This is a ghostly apparition, created when an enormous shadow of a person is cast on mist or clouds, often with a dramatic rainbow ring, known as a glory, around the shadow's head. In 1894, the Scottish physicist Charles Wilson was working on the summit of Scotland's highest mountain, Ben Nevis and saw a Brocken specter. This inspired his interest in the physics of clouds.

This dramatic Brocken specter image on the clouds appears extremely large due to an optical illusion—the cloud is considerably closer than it appears to be.

As investigating clouds was not Wilson's main occupation, it was another 17 years before Wilson had invented a device that successfully recreated clouds in the laboratory. These were located in small chambers, and by the time Wilson had created his first successful device in 1911, the subject of his study had shifted from clouds themselves to so-called ionizing radiation.

Ionizing radiation is the dangerous type of radiation. All light is radiation, but it is only the high-energy forms of light—gamma rays, X-rays, and some ultraviolet—that are ionizing. Similarly, high-energy matter particles—most often alpha particles (helium nuclei), beta particles (electrons) and neutrons—are ionizing radiation. The "ionizing" part implies that this radiation is energetic enough to create ions, which are atoms that have gained or lost electrons. The radiation hits atoms and gives outer electrons sufficient energy to fly off.

Ionizing radiation causes cancer, radiation sickness, and death. It's the radiation that we all fear, whether from accidents in nuclear power plants and atomic bomb fallout, or too much exposure to strong sunlight. However, it's also the way that we can better study the particles that make up everything around us. Wilson noticed that when he had produced small artificial clouds, ionizing radiation would disrupt the even spread of water vapor in the chamber.

Wilson's cloud chamber worked by filling a sealed container with warm, humid air, then expanding the volume of the chamber with a diaphragm. The result was the formation of a cloud, essentially an aerosol of tiny water droplets that are too small to see individually. When a particle of ionizing radiation passes through the chamber, it leaves behind a trail of charged particles, which create natural focal points for significantly larger water droplets to form. In effect, the charged particles seed the cloud in the chamber, producing a track of visible droplets.

Wilson's chamber would be improved upon in 1936 by American physicist Alexander Langsdorf Jr., whose diffusion cloud chamber was both more sensitive and longer lasting than Wilson's. The diffusion cloud chamber constantly introduces heated alcohol vapor into the top of the chamber, which is cooled from the bottom, making it capable of longer-term and more effective observations.

ANTIMATTER AND MUONS

The two patterns of the greatest importance discovered in cloud chambers were both the work of the American physicist Carl Anderson. The first of these involved a substance that has become a science fiction staple—antimatter.

The existence of antimatter had been suggested in the late 1920s by the English physicist Paul Dirac. In combining quantum physics and the special theory of relativity, Dirac discovered that there ought to be a particle that was, in effect, the absence of an electron with negative energy. This should be identical to the more likely possibility of the presence of an equivalent to the electron that had positive energy and a positive charge.

Ironically, several years later, Dirac, who had spent almost his entire life at Cambridge was in America on sabbatical at Princeton University when the American physicist Robert Millikan gave a presentation in Cambridge on a discovery made by Millikan's graduate student, Carl Anderson, that involved using a cloud chamber on cosmic rays.

ANTIMATTER: This is created when energy is converted into mass. Matter particles are created in pairs, one matter, one antimatter.

The way that cloud chambers were used to study particles was by looking at how the path of a particle curved when it passed through a strong magnetic field. If the particle had no electrical charge, the pattern of water droplets it left behind would travel in a straight line. But if the particle was charged, it would curve off in a direction that was determined by whether its charge was positive or negative, with a curvature that reflected the strength of its charge and its momentum (combined mass and velocity).

What Anderson had discovered was a particle with the same charge and mass as an electron, but with a positive charge, not a negative one. When energy, such as that of the incoming cosmic ray particles, is converted into matter, what happens is that a pair of particles is produced. Electrical charge is conserved—it is neither lost nor gained. As a result, when a pair of charged particles are produced, one is positively charged and the other negative. The electron is the negative part of the pair. The positive one is its antimatter equivalent, an antielectron, better known as a positron.

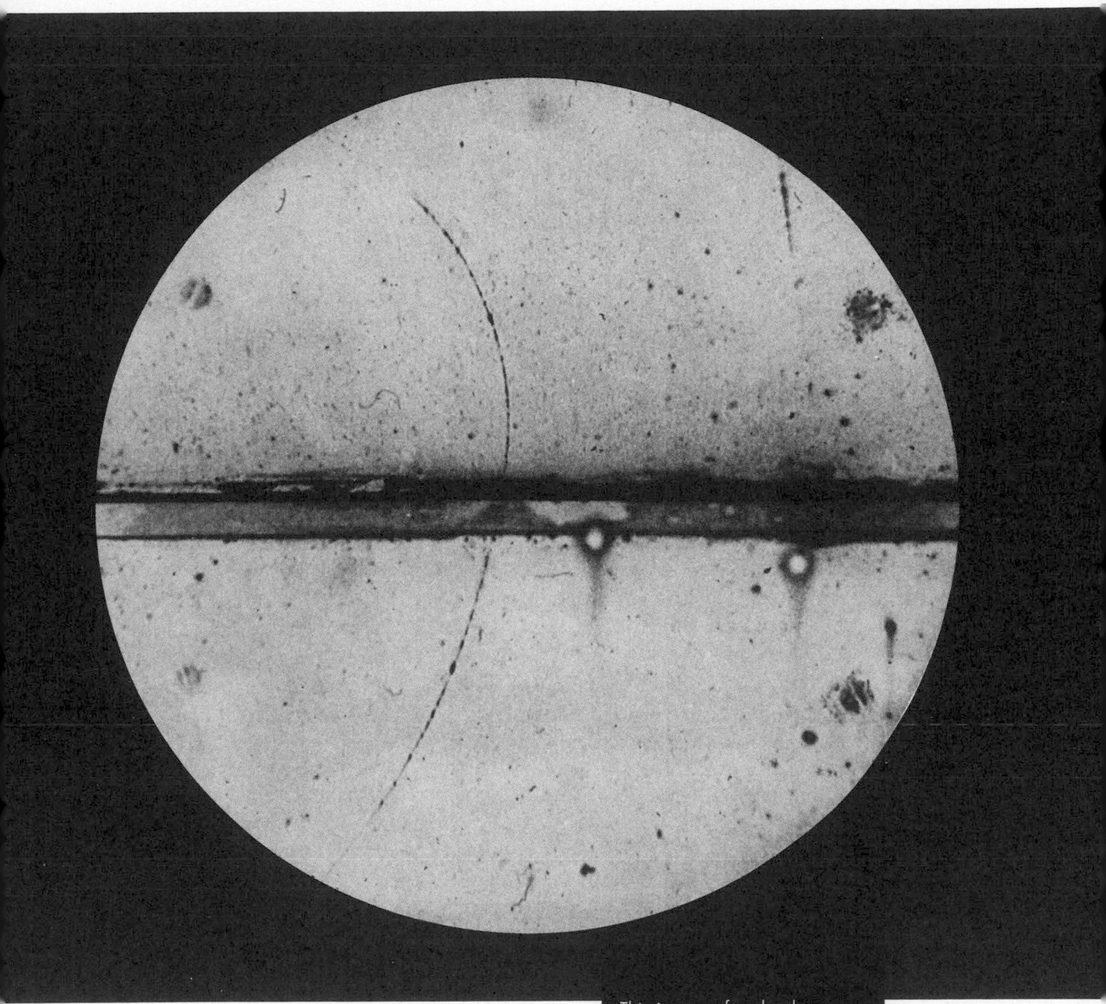

This image of a cloud chamber was taken by Carl Anderson. It shows the creation of an electron and an antimatter positron, which follow paths that curve in opposite directions.

This was the first dramatic breakthrough made in the patterns of a cloud chamber. Carl Anderson would also discover the muon as a product of cosmic ray radiation in 1936. In 1947, another new particle, the kaon, which is one of a family of particles known as mesons, was also discovered in a cloud chamber from a cosmic ray shower, this time by English physicists George Rochester and Clifford Butler. But the reign of the cloud chamber was soon to come to an end, replaced by a more liquid approach.

FROM BUBBLES TO WIRES

As we have seen, cloud chambers relied on having the saturated gas of a substance, such as water or alcohol, ready to form tiny droplets. There are some limits to the stability of this setup. The replacement was the bubble chamber, which uses a transparent liquid (often liquid hydrogen), heated above its boiling point, but kept liquid under pressure. This state is known as being superheated.

When an ionizing particle passes through the liquid, the liquid vaporizes, leaving behind a trail of bubbles. Bubble chambers tended to be larger than cloud chambers and produced better resolution. The device was invented by the American physicist Donald Glaser in 1952.

NOBEL PRIZE: The 1960 Nobel Prize in Physics was awarded to Donald Arthur Glaser for the invention of the bubble chamber.

The most successful bubble chambers were used at CERN, the European particle research center based near Geneva and sited on the border of Switzerland and France. Their discoveries mostly involved detecting neutrinos, produced as a result of interactions of the weak nuclear force, and the particles that carry that force, listed as bosons on the standard model diagram—W and Z particles.

The star here was a chamber known as Gargamelle. This 15.7-feet (4.8-meters) long chamber, 6.5 feet (2 meters) in diameter, held liquid freon, a refrigerant liquid based on chloro- and fluorocarbons. The name reflected the large size of the chamber. (Gargamelle was the fictional mother of the giant Gargantua in a sixteenth-century novel by French writer François Rabelais.) Gargamelle had a working life of just nine years before a crack rendered it beyond repair. Even if it had been possible to continue using it, though, bubble chambers were already obsolete.

Modern detectors are variants on the concept of the wire chamber. This features a grid of electrically charged wires in a chamber where the walls have the opposite electrical charge. When an ionizing particle passes through the chamber it ionizes the gas in the chamber (often a combination of a noble gas such as argon and an organic gas such as butane). Ionized substances conduct electricity. This is the reason, for example, that water is a conductor. Totally pure water doesn't conduct electricity—it is the charged particles usually dissolved in water that do.

An image showing particle creation in a bubble chamber at the Stanford Linear Accelerator Center.

Air is usually an electrical insulator, but it becomes a conductor when ions allow electrical energy to flow as lightning bolts.

Similarly, when lightning tears through the sky, it first ionizes the air before the main bolt of electricity can pass through. Within the wire chamber, the ionization produced by the passing particle allows a brief electrical current to flow, which is detected at multiple locations, pinning down the path of the particle.

The reason that variants of this kind of chamber have taken over from their bubbly predecessors is rather similar to the reason that professional astronomers no longer spend time looking through telescopes. It is much more efficient to have the data captured by a series of detectors, enabling it to be handled immediately by a computer. Patterns in the old types of chamber had first to be photographed. The photographs would then be worked over by researchers, laboriously making careful measurements of each track and calculating the outcome. In a modern chamber, the patterns are built up directly in a computer.

INSIDE THE WORLD'S BIGGEST MACHINE

In truth, it's impossible to imagine that the most dramatic producer of these patterns, the Large Hadron Collider (LHC) at CERN could ever have functioned with one of the old chambers—and certainly there would have been no chance of making a sighting using one of the LHC's great discovery, the Higgs boson.

The LHC is a thing of superlatives. It has been described as the largest machine in the world. When academic papers are published on its work, the total number of authors can be immense, stretching into the thousands. It consists of a vast ring 17 miles (27 km) long with a number of large detectors at various points around it, notably ATLAS (A Toroidal LHC Apparatus) and CMS (Compact Muon Solenoid). Despite that "compact" is in the name, CMS weighs in at around 15,432 tons (14,000 tonnes), and measures 69 feet (21 meters) long and 46 feet (14 meters) in diameter.

The LHC is more generally known as a particle accelerator. Initially, such devices were desktop machines, like the cyclotron built in 1931. But to get greater and greater energies, accelerators have grown through the decades, culminating (to date) in the Large Hadron Collider. Although the technical details are exquisitely complex, the basic concept is almost prehistoric in its basis. An accelerator makes particles move very quickly, then smashes them together to see what happens. If there is sufficient energy in the collision, new particles will be generated.

This is exactly the same as the collisions that occur when cosmic rays enter our atmosphere, but even in the LHC, the particles crash into each other with far less energy. It might seem pointless to spend so much money (around $5 billion in the case of the LHC) when there is a better natural source. But, the big advantage of using an accelerator is that the stream of particles is controlled and much easier to monitor.

The LHC made use of an existing tunnel built for an earlier accelerator. It uses huge superconducting magnets to accelerate protons (the "hadrons" in the name—hadrons are protons or neutrons) up to around 99.999999 percent of the speed of light. The particles flash around the accelerator in beams traveling in opposite directions before the beams are brought head to head in the detectors. At this point, the energy of the collision is converted into new particles.

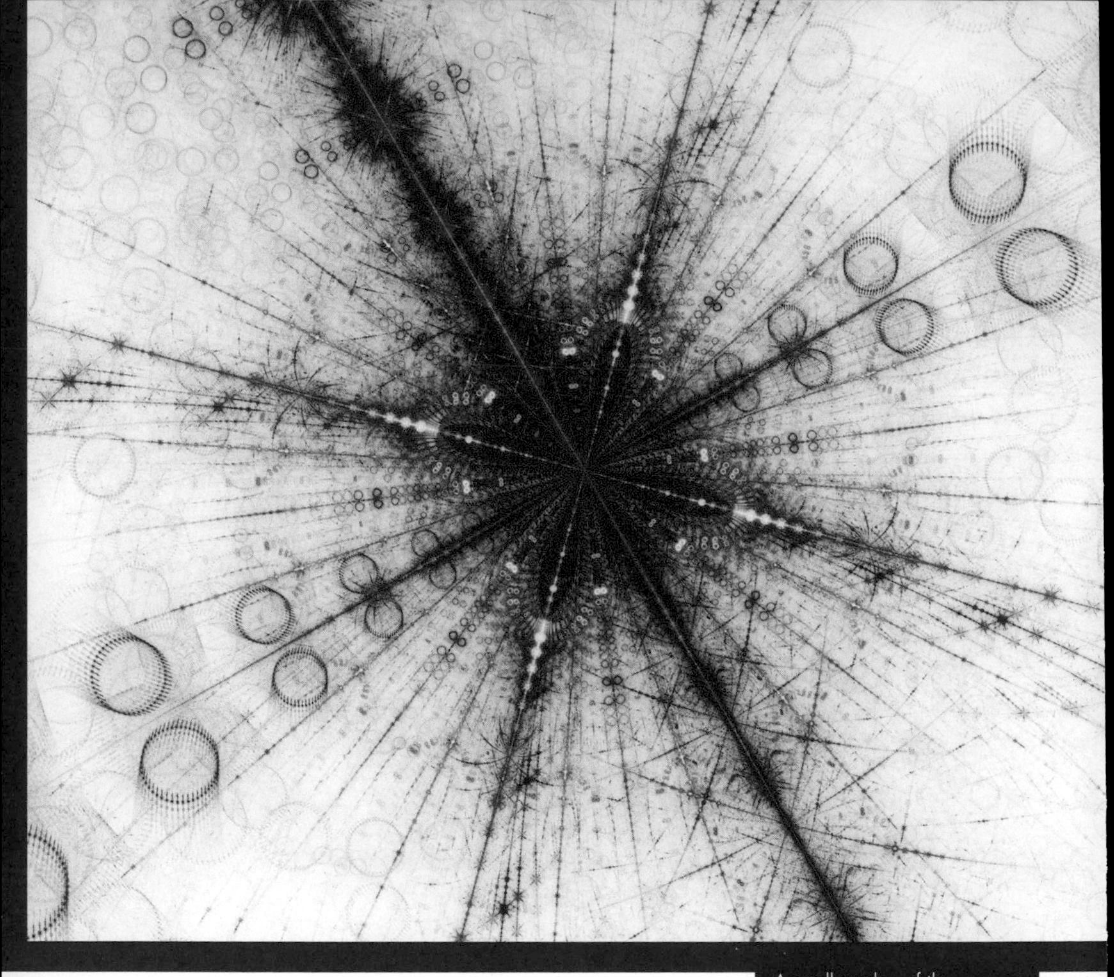

A small number of the particles produced in a typical event in a collider.

When the positron was discovered, the particle pattern was simple and clean. The LHC's detectors analyze vast explosions of particles, each of which may then create its own shower of new particles, requiring a huge volume of data to be captured and sifted through. The scale is phenomenal. During a run there are around 600 million events a second, requiring around 25 gigabytes of data per second to be stored. This is beyond the capability of even CERN's phenomenal data-processing capacity.

Instead, a set of algorithms is designed to weed out the majority of the uninteresting data and throw it away. Those 600 million events in each second are reduced to 100,000, and then on a second pass to between 100 and 200 events per second. This still results in a huge amount of data. This can be seen in the way that it took a good two years from first collecting data on the Higgs boson in 2010 to the announcement of its discovery in 2012.

FINDING THE HIGGS

No one has seen a Higgs boson. No one even saw the familiar kind of pattern created in an old-fashioned cloud or bubble chamber. Although we can look at elegant looking patterns as representative of the collisions that resulted in the Higgs being discovered, they were produced after the fact. Modern detectors deal in mathematical patterns in the data. It's a whole different world of pattern detection and understanding. But it was enough for physicists to put the chances of the detection they made occurring without the Higgs existing at less than one in a million.

The detection of the Higgs boson has been by far the most dramatic discovery made by the LHC. The Higgs apart, most of the LHC's explorations have been negative. This isn't as bad as it sounds. Although the basic standard model of particle physics seems very robust, it can't explain everything that is observed. Theoretical physicists have postulated for decades that there might be as many particles again, each a "supersymmetric partner" of one of the existing particles; each matter particle (fermion) would have a partner force particle (boson) and vice versa. So, for example, the photon, which is a boson, would have a partner called a photino that was a fermion, while a quark (fermion) would have a boson equivalent called a squark.

Theory suggested that if these particles existed, some should be within the reach of the LHC because it should have had enough energy to produce them. As yet, though, not a single supersymmetric particle has been produced, significantly reducing the possibility that supersymmetry theory is correct. This is a negative result that has positive value. In the case of the Higgs boson, though, we had a concept that dated back theoretically to the 1960s, which was finally verified in 2012. When the discovery of the Higgs boson was announced there was much coverage in the press, but little clarity over what the Higgs boson actually was. It was often represented as "the particle that gives all the other particles their mass." In reality, that statement is almost entirely incorrect.

Firstly, the most familiar particles with mass are electrons, protons, and neutrons. Of these, both the proton and the neutron get almost all their mass from another source. Each is composed of three fundamental particles called quarks—and quarks are bound together by a large amount of energy. So much so, that it is almost

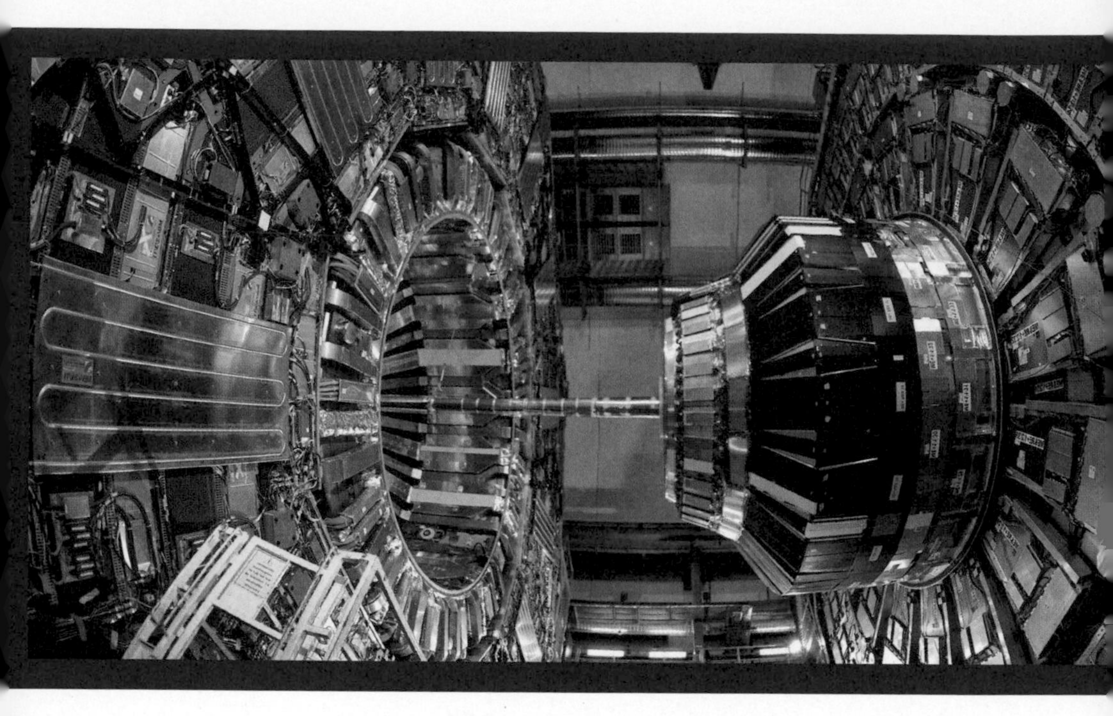

impossible to reduce a proton or a neutron to its constituent quarks. But, as we have seen, energy and mass are interchangeable. If something has energy, that energy has mass—and almost all of the mass of these particles comes from the energy that binds the quarks together.

Most particles that aren't made up of quarks would be expected by theory not to have mass. The theory of where it comes from postulates that the universe is filled with a field called the Higgs field. Remember that many particles can be represented as moving spikes in a field—something that has a value throughout space and time. The Higgs field is thought to interact with other particles, making them behave as if they have mass.

The Higgs boson is a disturbance in the Higgs field, just as a photon can be considered a disruption in an electromagnetic field. It is not the Higgs boson that gives particles their mass, it is the Higgs field. The existence of that field was discovered by spotting the boson when a spray of particles was seen coming into existence, because the Higgs boson would not have stayed in existence long enough to be detected. That spray provided a signature that convinced scientists the Higgs boson had really been there.

Left: Compact Muon Solenoid, one of the huge particle detector experiments that is part of the Large Hadron Collider at CERN.

Right: An artist's impression of a Large Hadron Collider event showing the tracks produced by a decaying Higgs particle.

The patterns from particle colliders represent the actual interaction of individual particles, however big and complex those colliders have become. But quantum physics, the physics of the very small, including these particles, is distinctly deceptive. In talking about particles so far, we have acted as if quantum particles were tennis balls with predictable paths that could be measured.

In reality, the pattern from a collider is the outcome of a far more complex quantum behavior that involves the probability of the particle taking every possible path that is available. Remarkably, these probability patterns have proved essential in the development of quantum theory, where they are represented by Feynman diagrams.

> **"I AM QUITE SURPRISED THAT IT HAPPENED DURING MY LIFETIME. IT IS NICE TO BE RIGHT ABOUT SOMETHING SOMETIMES.** PETER HIGGS

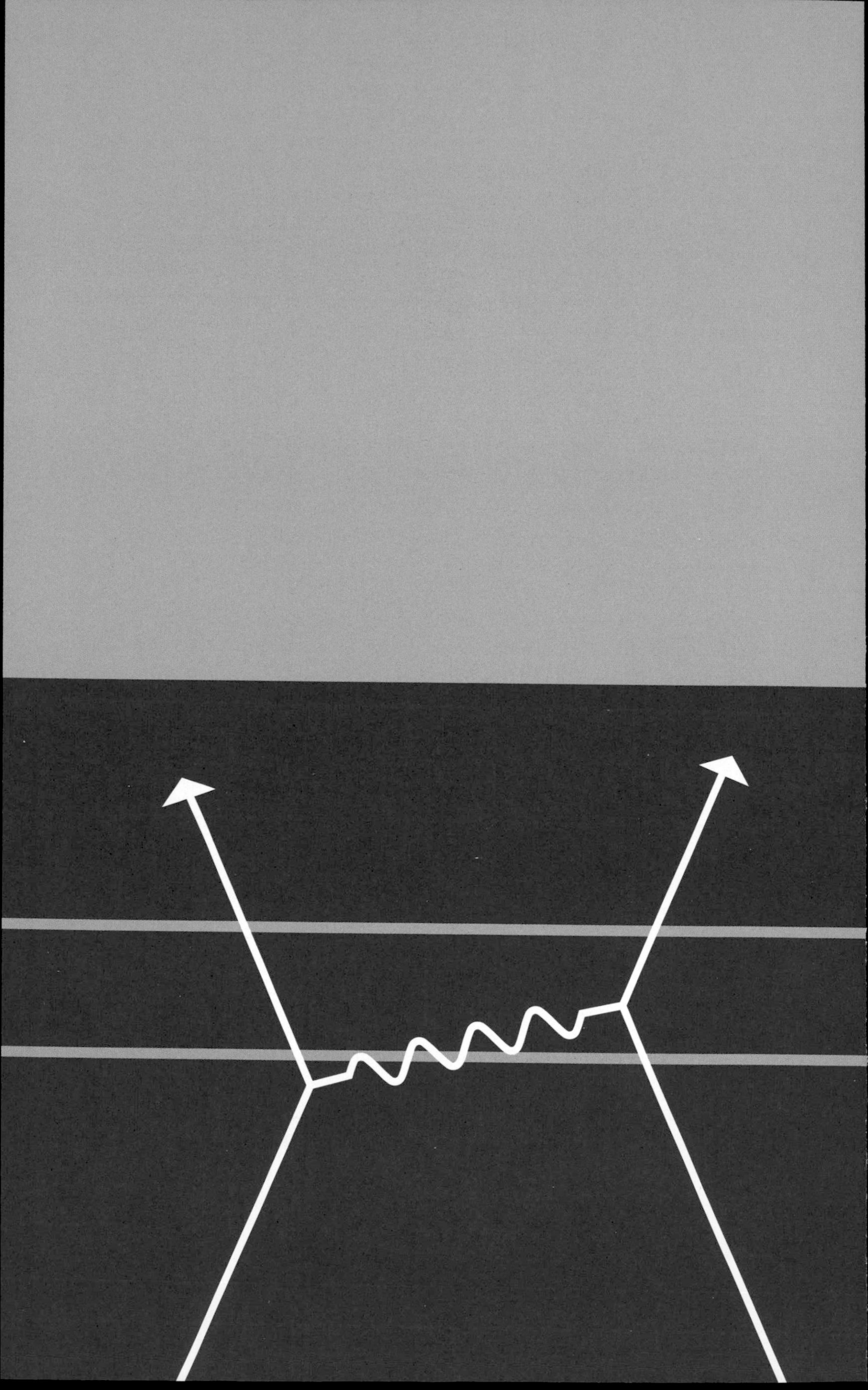

4
FEYNMAN DIAGRAMS

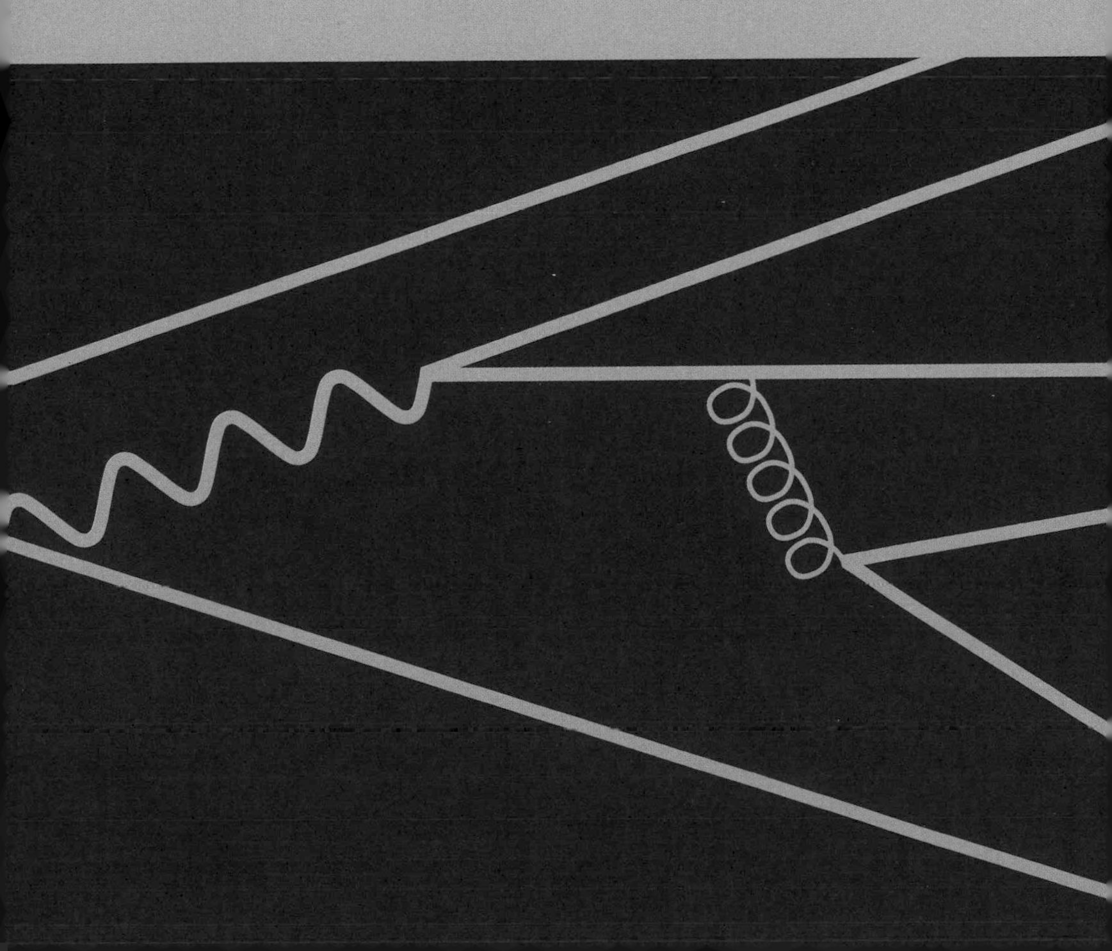

Richard **Feynman**
1918–1988

LIGHT AND MATTER

Quantum physics describes the workings of the universe at the level of the tiny particles, such as electrons and photons of light, that make it up. Most of our everyday experiences come from the interaction of matter particles with each other and with light, which can be explained by quantum electrodynamics (QED). This detailed theory would win the Nobel Prize in Physics for Richard Feynman, Julian Schwinger, and Sin-Itiro Tomonaga. Much of their original thinking involved complex mathematics, as it was necessary to deal with a vast number of potential interactions, each with different probabilities. But Feynman realized that these interactions could be represented by simple diagrams—patterns that define the interplay of light and matter. Not only did they make QED more comprehensible, Feynman diagrams also provided a visual tool for making otherwise impossible calculations practical.

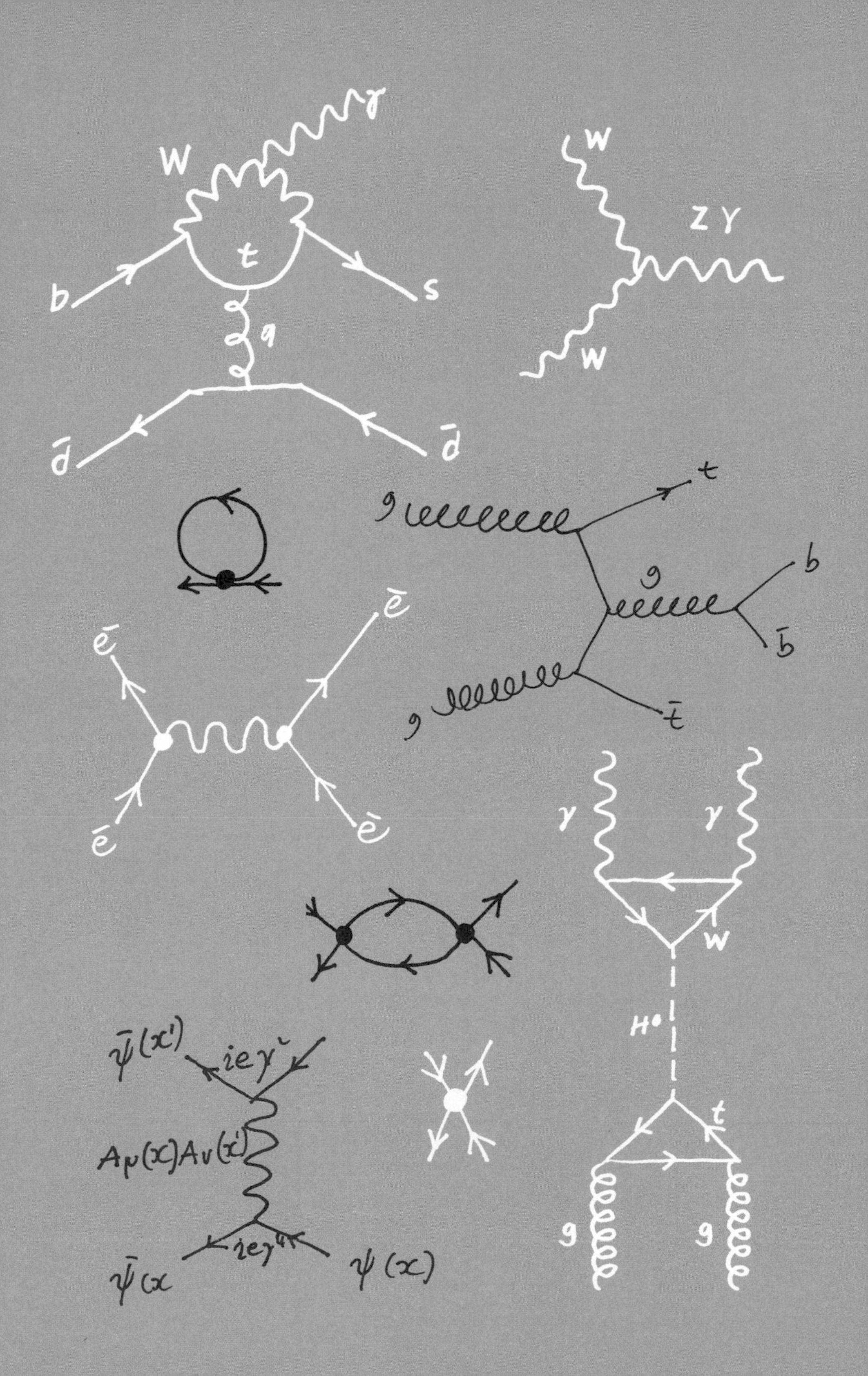

γ
W
t
b
s
g
d̄
d̄
W
Z γ
W
t
g
g
b
b̄
g
t̄
ē
ē
ē
ē
γ
γ
W
H°
t
g
g
ψ̄(x')
ieγ^ν
Aμ(x)Aν(x')
ieγ^μ
ψ̄(x
ψ(x)

Richard Feynman was so proud of his diagrams that he had his 1975 Dodge Tradesman Maxivan—license plate QANTUM—painted with them, becoming a familiar sight on campus.

THE PHYSICIST'S PHYSICIST

Feynman diagrams help us to explore how matter and light particles interact and make complex calculations more practical. But to understand the diagrams, we need first to understand the man.

Richard Phillips Feynman, who died in 1988, is still revered among physicists. In part this is due to his capability and range of imagination, but it is also thanks to his direct and insightful ability to communicate, which tends to be rare among the physics fraternity. A dramatic example arose when Feynman was asked to join the commission set up to investigate the 1986 Challenger space shuttle disaster. He was reluctant to take part, but his wife, Gweneth, encouraged him to do so; she felt he would be able to cut through the bureaucracy and uncover a cause that might otherwise be suppressed by vested interests. Feynman did just that, escaping the highly controlled fact-finding visits and doing his own research.

Discovering that the O-rings that sealed joints in the rocket motors were likely to lose flexibility at low temperatures, Feynman was not willing to wait for the bureaucratic wheels to grind slowly on. He preempted a televised session of the commission, dunking a section of O-ring into his iced water and demonstrating on camera how slow it was to recover its shape. It was this same ability to communicate and his childlike directness that led Feynman to develop his iconic diagrams.

FORCE CARRIER: The two negatively charged electrons exchange a photon (a light particle) that transfers the energy necessary for the particles to repel each other.

Each diagram combines a series of lines where, for example, straight lines represent matter particles and wavy lines are photons. Like Minkowski diagrams these are patterns in space-time, but they respent the interaction of particles. A simple example might show two electrons repelling each other electromagnetically, with a photon passing between them as a force carrier. Feynman's diagrams were necessary to reflect the strange behavior of quantum particles, which bears no resemblance to the action of the physical objects that are made up of them.

To see why the diagrams were so significant, we need to take a step back to what makes quantum physics appear so bizarre.

"THE THEORY OF QUANTUM ELECTRODYNAMICS DESCRIBES NATURE AS ABSURD FROM THE POINT OF VIEW OF COMMON SENSE.

RICHARD FEYNMAN

QED: QUANTUM ELECTRODYNAMICS

Quantum physics began with Einstein's realization that photons were real, soon expanding to explain the structure of the atom. It is the science of the very small, where reality seems not to have the deterministic certainty of the world we typically observe, but rather probabilities dominate. Quantum electrodynamics, which is where Feynman diagrams were first used, considers the interactions of quantum particles that are dependent on electromagnetism.

QED: Quantum electrodynamics explains the electromagnetic interactions of quantum particles.

In common usage "electromagnetism" sounds like it is just about electricity and magnetism, which in a sense it is. But we need to understand that it is responsible for the vast bulk of everyday interactions in the world that we experience. Light is an electromagnetic phenomenon. Similarly, most interactions between atoms is electromagnetic. So, for example, when you sit on a chair it is the electromagnetic force acting between the atoms of the chair and the atoms in your body that prevent your atoms from simply slipping past those in the chair.

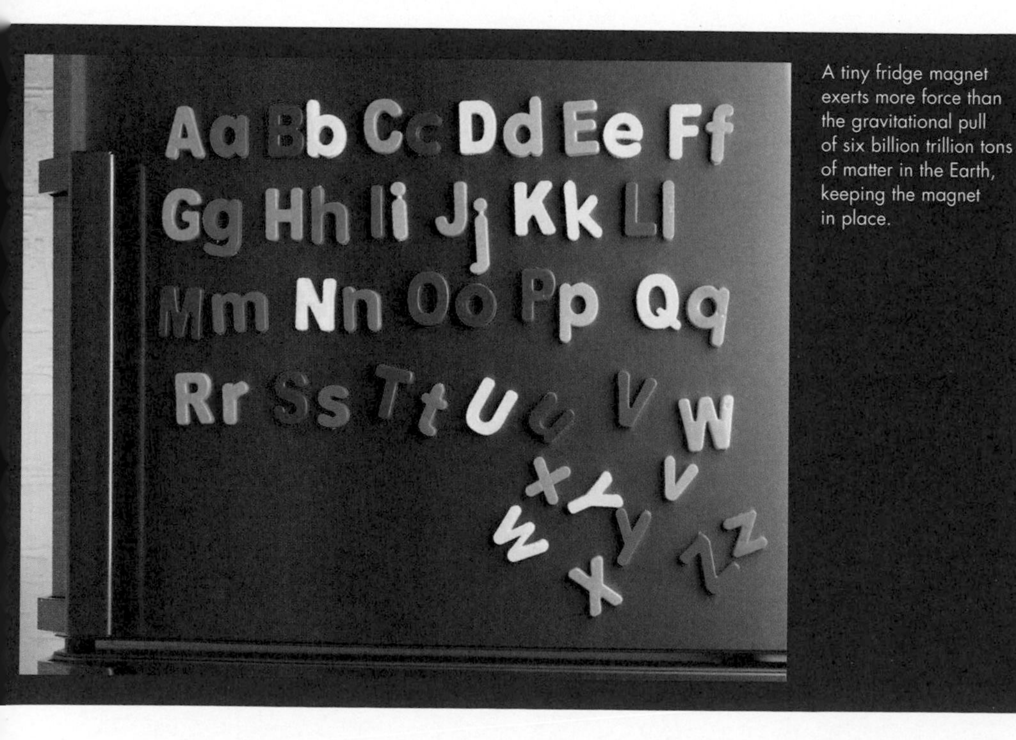

A tiny fridge magnet exerts more force than the gravitational pull of six billion trillion tons of matter in the Earth, keeping the magnet in place.

The electromagnetic force is one of the four fundamental forces of the universe. The other three are gravity, and the strong and weak interactions—these two forces are involved in the nucleus of atoms. We tend to think of gravity as pretty overwhelming, yet in fact it is by far the weakest of the four forces, billions of billions times weaker than electromagnetism. If you doubt this, just think of a fridge magnet. The entire gravitational pull of Earth is attempting to pull it to the floor. All that holds it to the fridge is the electromagnetic force from its tiny magnet. The magnet wins.

In quantum theory, forces pass from place to place as a result of so-called force carriers—particles that travel between the two objects that are attracting or repelling each other. This is why, for example, a magnet can attract a piece of iron at a distance. Perhaps surprisingly, the force carrier of electromagnetism is a particle we have already met—the photon.

We usually get introduced to the photon as the particle of light, but every time there is an electromagnetic interaction between matter particles a flow of "virtual photons" between the particles produce the effect of a force. The term "virtual" here is decidedly misleading. It sounds as if it means that the particles don't exist. What it really refers to, though, is that the photons aren't ever observed, as they pass from one particle to another without escaping.

As a result, nearly every electromagnetic interaction—which means nearly every interaction of matter not involving gravity—is the result of a matter particle emitting a photon, or a matter particle absorbing a photon, or both.

“THE QUANTUM THEORY OF THE INTERACTION OF LIGHT AND MATTER, WHICH IS CALLED BY THE HORRIBLE NAME ‘QUANTUM ELECTRODYNAMICS.’

RICHARD FEYNMAN

THE MAGIC MIRROR

Before seeing how Feynman diagrams help understand what is happening in a quantum physics interaction, it's worth looking at a simple example of why things are far different from expectation when the universe is operating at the level of quantum particles. Think of an example from the kind of optical physics we might learn at school: a beam of light hits a mirror and bounces off. We are taught that "the angle of incidence is equal to the angle of reflection." In other words, the light bounces off at the same angle to the mirror as it arrives. This is what we would expect if the photons in the light behaved like little tennis balls. Unfortunately, the picture bears no resemblance to what actually happens. What we're taught at school is entirely wrong.

Quantum reflection

We usually expect a reflection to happen at the same angle as the light arrives at the mirror. But when strips are removed, light reflects at an unexpected angle.

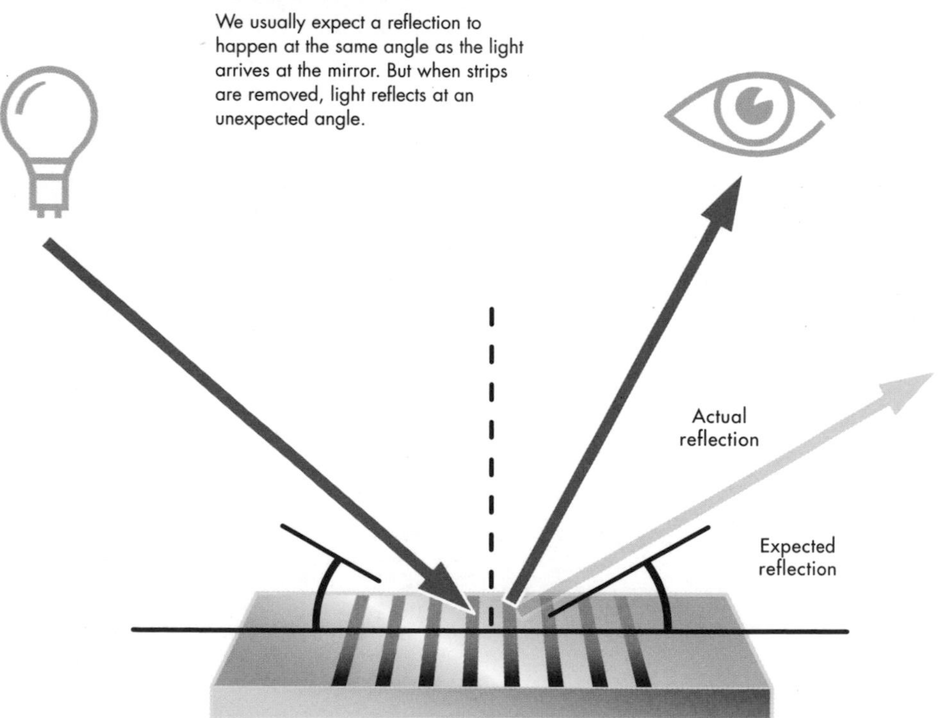

Mirror with strips missing

Instead, quantum particles don't travel on neat straight-line paths. They get from A to B somehow, but rather than having a specific trajectory, they have differing probabilities of taking every possible path between A and B. In effect, they don't take any specific path, unless we force them to do so by observing them in flight. As it happens, though, various different paths cancel each other out so that the outcome is as if they did indeed take that familiar equal angle bounce. But if the outcome is the same, how can we say that the traditional description is wrong? Because we can remove some of those canceling probabilities and get the light to go in a totally different direction.

PROBABILITY RULES: In quantum physics, particles don't take specific paths, but have combined probabilities of taking many different paths. It was quantum theory's probabilistic nature that Einstein disliked.

If strips of the mirror are removed many of the canceling paths are impossible to take. The result is that the light reflects at an unexpected angle. It's a bit fiddly to make a mirror with these missing strips as they have to be quite narrow, but a familiar object does this for us: a CD or DVD. Under the plastic surface of these optical discs is a reflective surface that has little pits where the reflective part is missing. This is how the CD stores information. Tilt the CD at an angle and you will see rainbow-like colors in the reflection. This is where the pits are cutting out some of the possible reflection angles and the light will reflect at unexpected angles. As this effect is dependent on the energy of the photons, different colors reflect at different angles, causing the colored pattern.

" I, AT ANY RATE, AM CONVINCED THAT [GOD] IS NOT PLAYING AT DICE.

ALBERT EINSTEIN

PARTICLE PLAYTIME

Feynman diagrams are designed as a way both to illustrate those electromagnetic interactions, and to explore and quantify the many variants that the strange nature of quantum physics provides that would not be expected otherwise. On the diagrams, matter particles are represented by straight lines and photons as wiggly lines. (There are other types of line when the use of the diagram is expanded beyond simple QED.) Unlike the Minkowski diagram (see page 40) there is not such a clear convention on which axis is time and space. Quite often time is the vertical axis in a Feynman diagram as it is on the Minkowski diagram, but where it's more convenient, it can be the horizontal axis instead.

The most common elements shown on such a diagram are that a photon travels from one location to another; a matter particle (at its simplest, an electron) travels from one place to another; or a matter particle gives off or absorbs a photon. Almost everything can be built up from these simple components. But because of the oddities of quantum of quantum physics, an apparently simple action can result in a whole plethora of diagrams.

Take an apparently simple example of two electrons moving. We know where they start and where they finish. But how do they get from being located at A and B to being at C and D? The simplest possibility is the electron from A ends up at C, and that from B at D. Another probability is that the electron from A ends up at D and the one from B at C. Note that we can't tell which has happened as we don't know the route the electrons took, nor do we know which electron is which. One of the definitive things about quantum particles, such as electrons, is that they have no distinguishing features. They are truly identical.

That's straightforward enough, but there are other possibilities. Electrons, like other quantum particles, can undergo a process known as scattering. This is often represented as one electron bouncing off the other, like a pair of balls on a snooker table. However, electrons are electrically charged particles, and electromagnetic interaction works through photons. So, another diagram would show a photon passing from one electron to the other, as a result changing the paths of the electrons so that they end up at C and D. This could arise in a number of ways.

Each of the different possible diagrams will have a probability attached to it. As we add in more unlikely possibilities, the outcome becomes closer and closer to reality. It's amusing that quantum physics is in one sense the most accurate science we have. As Richard Feynman once observed, the difference between its predictions and reality are comparable to the width of a hair on the scale of the distance between New York and Los Angeles. Yet on the other hand, the predictions of quantum physics are based on probabilities, and though we can come nearer to the actual value it will always be the limit of taking every possible diagram into consideration, rather than a simple outcome.

Electron diagrams

Increasingly complex Feynman diagrams show just some of the ways two electrons can start from A and B and end up at C and D.

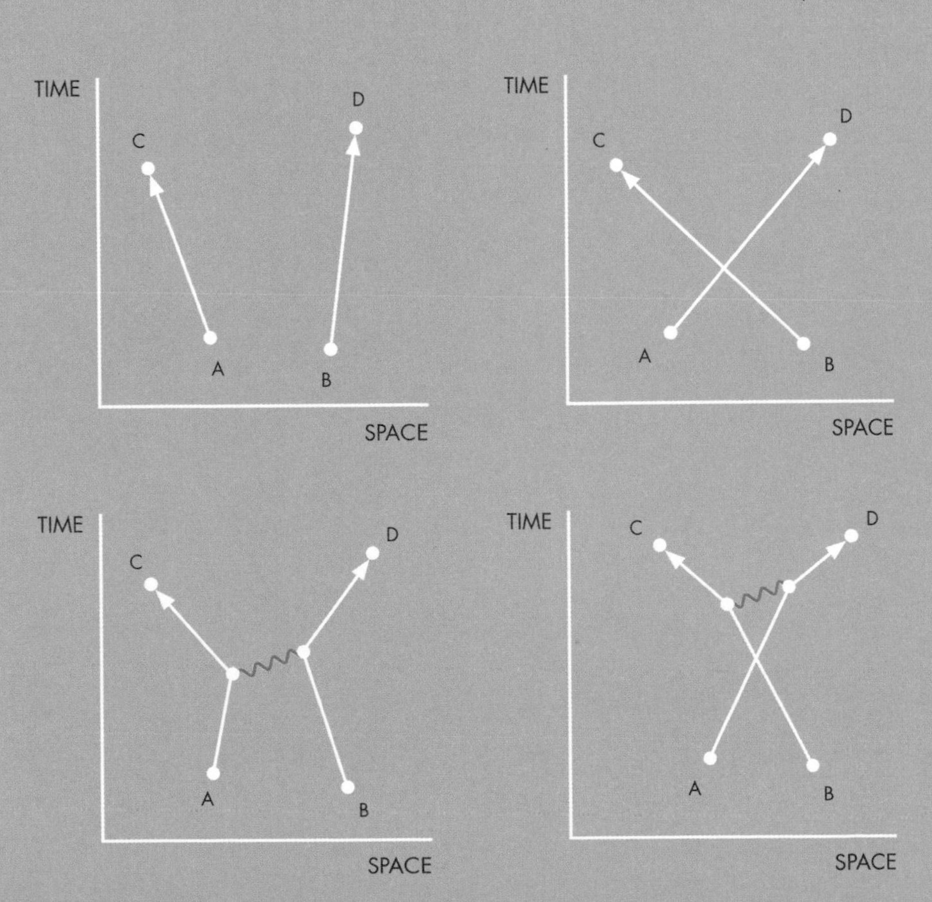

Although Feynman diagrams sometimes have arrows indicating a direction of travel, they are often not necessary. Take, for example, the photon traveling between electrons in the top two diagrams on page 85. It is perfectly acceptable to say that the photon travels in the direction that has it traveling into the future, but in practice, the mathematics used to make the calculation doesn't care whether the photon travels forward or backward in time. Just as the diagrams don't usually distinguish a direction, so the photon is described as being "exchanged" by the particles, rather than traveling from a specific particle to the other one.

We won't go through every possible diagram for this simple interaction (in fact it would be impossible, and even attempting it would be highly tedious). But just to show how more and more complexity could be added in, a next possibility could be to have two photons exchanged along the way, producing a pair of scattering events. By the time we get down to an event like this, the contribution is already something like one part in 10,000. Note, incidentally, that the diagrams are more than visual illustrations—they are a mechanism to base calculations on. The actual calculations can become painfully messy, but the diagrams provide the patterns to ground them in an approachable fashion.

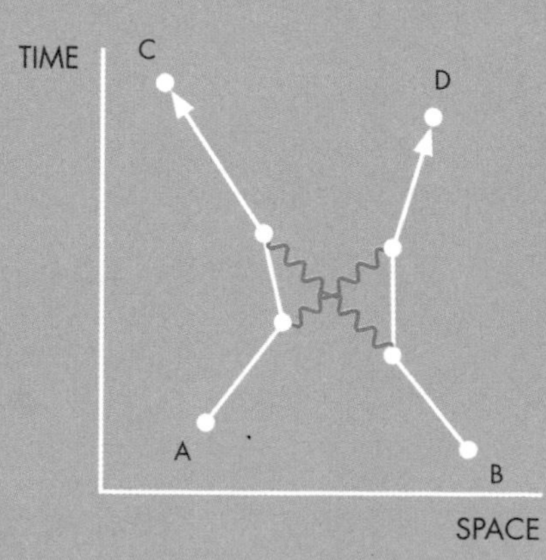

Increasing complexity

This shows the next level of complexity from the diagrams on the previous pages, the two electrons can interact by exchanging a pair of photons in a dual scattering event.

Sunlight is scattered by the interaction between photons and electrons orbiting atmospheric atoms, spreading blue light across the sky.

WHY THE SKY IS BLUE

Quantum physics explains the vast majority of our experiences, including many that seem quite mysterious: for example, why the sky is blue. A clue to this is the color of the Sun. We talk about daylight being "white light," but anyone drawing a picture of the Sun tends to color it in yellow, or even red if it's near sunset. In fact, the Sun's light is white, composed of a rainbow mix of colors. When that light reaches our atmosphere, some of the photons interact with electrons in the gasses that make up the atmosphere, undergoing scattering.

This means that a photon is absorbed by an electron, which increases in energy, before dropping back down in energy and giving off another photon. This process happens more with the high-energy photons at the blue end of the spectrum; so blue light is scattered more across the sky, leaving an increased yellow tint to the Sun. Toward sunset, the sunlight is passing through more of the atmosphere and is scattered furthered, leaving a red-looking Sun.

That scattering process is easy enough to represent in a Feynman diagram, but once again the possibilities are varied and uncover something particularly strange. Let's look at three of the ways that scattering can occur.

In the first diagram (below), an electron absorbs a photon, increasing in energy. After a little time, the electron reemits a photon, dropping back in energy. This is the kind of process we might imagine being the case in the atmospheric scattering that makes the sky blue.

However, the second diagram shows us something equally possible. Here, the electron drops in energy, giving off a photon first, then absorbs a photon to continue with more energy. In both cases, the scattering is a matter of an electron absorbing and emitting a photon.

The third example seems at first glance to be much the same as the second. But remember that the vertical axis on the diagram is time. So, in the middle section, when the electron has emitted a photon, that electron travels backward in time to a point where it absorbed a photon, then carries on forward in time. This sounds counterintuitive, but there is nothing in the physics that prevents this.

It can help to make sense of that diagram to understand that an electron traveling backward in time would be the same, physically speaking, as a positively charged electron traveling forward in time. We know that all electrons have a negative charge, but as we have already discovered (see page 64), there is an antimatter equivalent of an electron called an antielectron or positron. If we consider that central line in the third diagram to be a positron, heading forward in time, suddenly the whole thing makes more sense.

Scattering diagrams

Three of the ways that a photon can be scattered by interacting with an electron. In the third example the incoming photon creates a matter/antimatter pair.

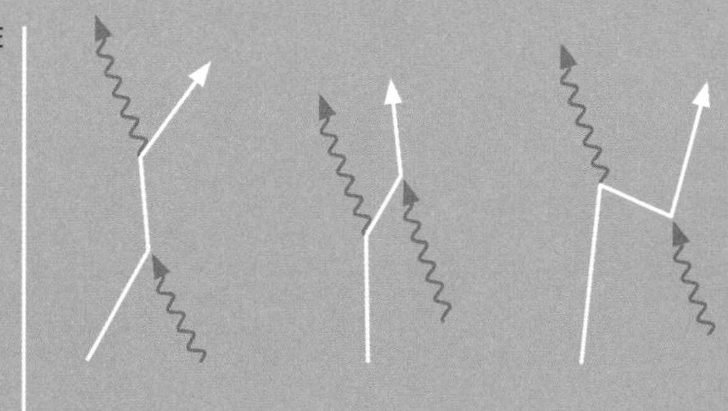

Now, what is happening in the right-hand part of the diagram is that a photon is converted into a pair of matter particles—a matter/antimatter pair—an electron and a positron. This is perfectly possible if the photon has sufficient energy according to $E=mc^2$ to make up the mass of the electron and positron. The electron flies off in one direction, while the positron heads off in another and meets up with an incoming electron. Now we get the opposite process known as annihilation. When an electron and positron meet, their matter is converted into pure energy—in this case in the form of a photon.

IT'S NOT JUST ELECTRONS

Although electromagnetism is primarily concerned with interactions involving electrons and photons, the Feynman diagram proved far too useful to limit it to these two types of particle, or for that matter to electromagnetism itself. One of the simpler examples would be the radioactive process known as beta decay. This involves another of the fundamental forces of nature, most closely tied to electromagnetism, known as the weak force or weak interaction.

In one form of beta decay, a neutron inside an atomic nucleus is converted into a proton. Because this changes the number of protons in the nucleus, which defines what the chemical element is, the result is transmutation. So, for example, carbon-14, the radioactive isotope of carbon used in carbon dating, undergoes beta-decay to become nitrogen-14. To get to the Feynman diagram for this process, we need to know a little more about what's going on inside neutrons and protons.

These relatively heavy particles were originally thought to be fundamental—not having smaller components. However, we now know that each is made up of three truly fundamental particles called quarks. Quarks come in six different "flavors," but only two are involved here—up and down. A neutron consists of two down quarks and one up quark, while a proton has two up quarks and one down. The beta decay process transforming a neutron into a proton results in a down quark becoming an up quark.

Where the electromagnetic force is carried by the massless, charge-free photon, the weak force has several variants of carrier particle, which have mass. We know that electrical charge is one of the aspects of the universe that is conserved: inside a closed system the charge cannot change. When a neutron changes to a positively charged proton (which could also be looked at as a negatively charged down quark changing to a positively charged up quark), negative charge is lost, so it has to appear somewhere else. In this case in the form of a weak force carrier particle, inelegantly referred to as a W-particle.

The W-particle is unstable and decays into two other particles—an antineutrino (as its name suggests this has no electrical charge) and an electron. This electron explains the term "beta decay," as before it was understood what the stream of electrons produced by this kind of nuclear reaction were, they were called "beta rays."

The existence of the neutrino was predicted before its place in the bigger picture was understood, because during such a reaction there is a loss of mass/energy, and something had to be carrying off that energy. However, neutrinos are extremely difficult to detect because they pass through most matter as if it were not there. Every second, billions of neutrinos from the Sun pass through your body unnoticed. As a result, it was a number of years from its theoretical prediction to the neutrino being first discovered.

We can now put together a Feynman diagram covering what happens during beta decay.

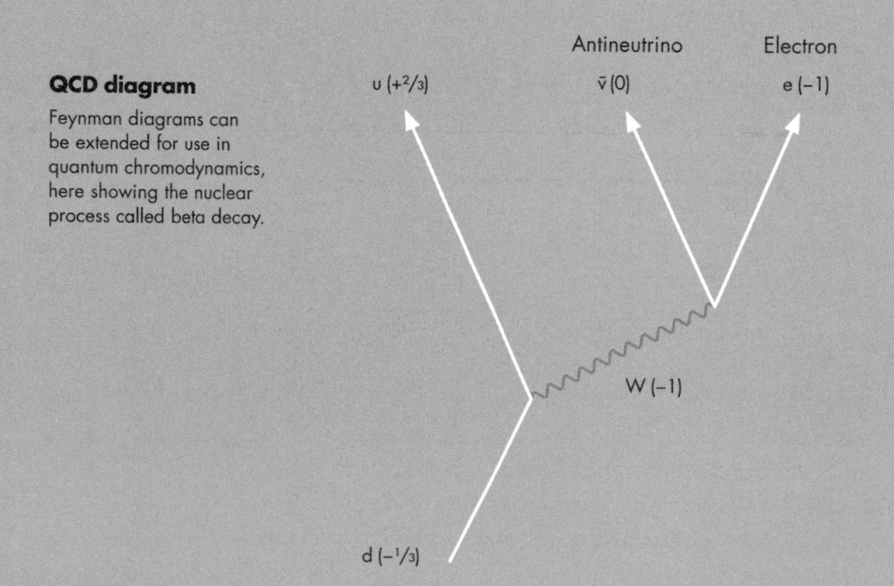

QCD diagram

Feynman diagrams can be extended for use in quantum chromodynamics, here showing the nuclear process called beta decay.

Anticolors

The "colors" in quantum chromodynamics correspond to the primary colors of light: red, blue, and green. "Anticolors" are the equivalent of the secondary colors: magenta, yellow, and cyan.

By bringing in reactions like beta decay, the Feynman diagram can cover the pattern of two of the four forces of nature. Gravity is yet to be covered by quantum theory, but the other fundamental force, the strong force, has its own equivalent of QED, known as QCD or quantum chromodynamics. Here the diagrams do exist, but can quickly become too complex to handle.

The strong force is the one that holds together quarks inside particles such as neutrons and protons, and leaks out sufficiently to hold together the atomic nucleus. It has its own force carrier, such as the photon without mass or electrical charge called the gluon. The quarks that gluons link together have both an electrical charge and another charge that has three different values. There is no suggestion that quarks come in different colors. But just as the three primary colors of light—red, blue, and green—combine to make white light, so the three color charges (also red, blue, and green) combine to make a neutral color charge.

In practice things get a lot more complex as gluons also have a color charge, or more precisely, in effect they have two color charges. Like other particles, quarks have antiparticles—and antiparticles have the opposite charge. Remember that the electron's antiparticle is the positively charge positron. But what is the opposite of red, blue, and green? If the physicists who came up with the naming convention had carried through their analogy, they should have gone for cyan, yellow, and magenta, which are the "complementary colors" to red, blue, and green. Instead, they unimaginatively went for antired, antiblue, and antigreen.

Each gluon effectively combines both a color and an anticolor, so it might seem that there should be nine different ones, but in practice these states themselves combine in different ways to produce eight different types of gluon. Because they have these charges, unlike photons, gluons can interact with each other. As a result, although the basic representation of strong force events on a Feynman diagram is perfectly possible, with gluons represented as a helix to distinguish them from photons, even a simple interaction can be messy. For instance, the simplest example is where an electron and positron combine to produce a boson that decays into a quark, an antiquark, and a pair of gluons. Bringing colors in makes the calculations more and more complex, requiring many thousands of diagrams to be considered.

QCD diagram

These simplified diagrams show some of the possible outcomes of an electron and positron annihilating each other.

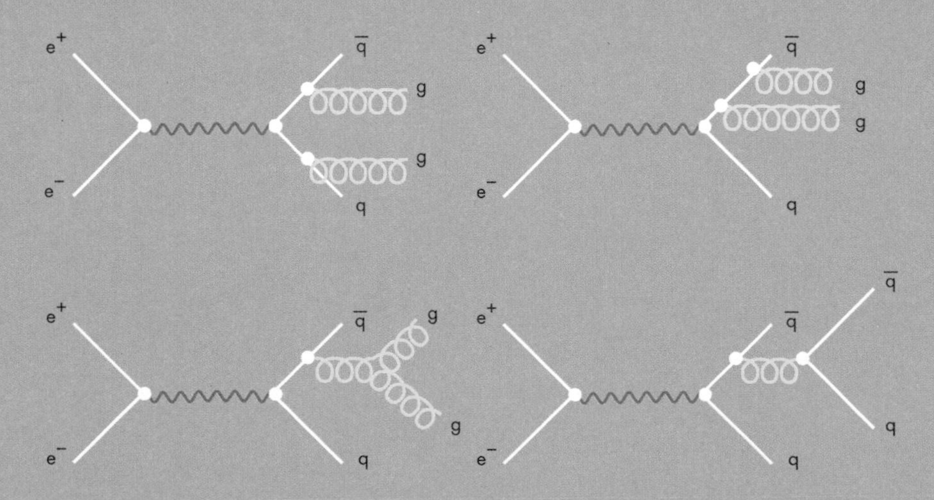

Positive Grassmannian

A multidimensional mathematical space can be represented by a diagram showing links between different dimensions, producing patterns such as this.

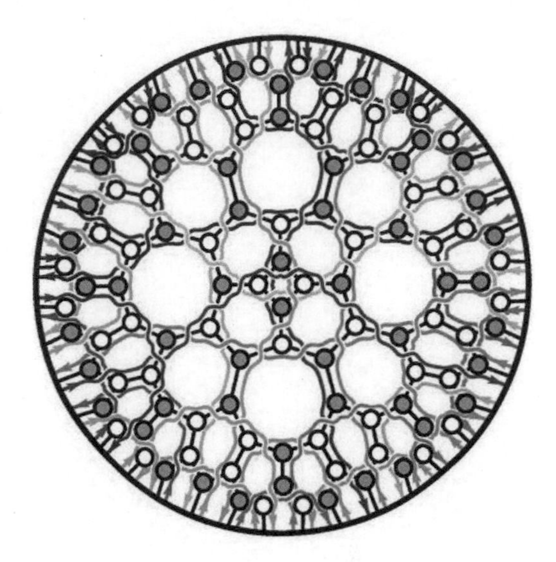

GOING BEYOND FEYNMAN

One possible future approach to the problem of providing an appropriate pattern for quantum chromodynamics is something known as an amplituhedron. This is based on a mathematical concept known as the positive Grassmannian, which is linked to the space inside a triangle. To form an amplituhedron, the concept of the positive Grassmannian is expanded into multiple dimensions, resulting in a kind of next-generation Feynman diagram that deals with many different particle interactions simultaneously.

An amplituhedron intertwines vast numbers of calculations into a single structure. The principle that these structures can be used to produce the outcome of these calculations is established. What is less clear is how to construct the specific amplituhedron for a chosen particle interaction in the first place.

Whether or not amplituhedra become practical tools, there is no doubt that Feynman diagrams have proved incredibly valuable in developing our understanding of interactions in the quantum world and carrying out the calculations necessary to make modern electronics possible.

The next pattern also reflects the outcome of quantum interactions, but ones that are less hidden than those studied in QED. This is the iconic pattern that lies behind much of chemistry: the periodic table of the elements.

5
THE PERIODIC TABLE

Dmitri **Mendeleev**
1834–1907

DEVELOPMENT OF THE PERIODIC TABLE

Chemical reactions are behind much of everyday life, from cookery to the workings of the body. One pattern sits at the heart of chemistry: the periodic table. The structure of the table reflects the underlying pattern of the configuration of atoms: each of the columns in the table corresponds to the number of electrons on the outside of the atomic structure. It is this pattern that determines how the elements will combine to make molecules, from simple compounds, such as sodium chloride, to magnificent natural edifices, such as the double spiral of DNA. Dmitri Mendeleev started to create what is now recognized as the modern periodic table. However, there were many other chemists whose work also contributed to the development of the periodic table.

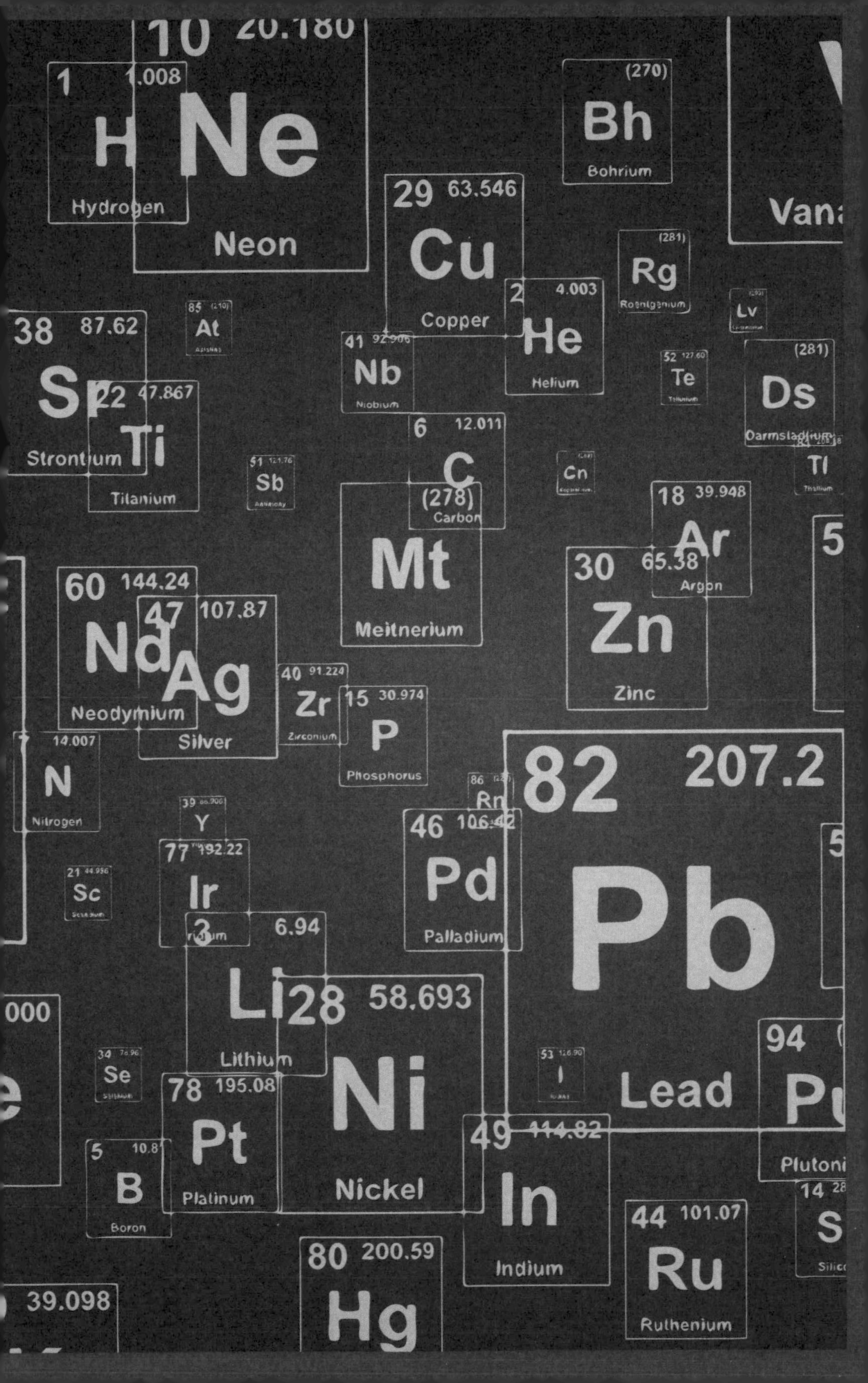

10 20.180
Ne
Neon
1 1.008
H
Hydrogen
(270)
Bh
Bohrium
29 63.546
Cu
Copper
(281)
Rg
Roentgenium
2 4.003
He
Helium
Lv
38 87.62
Sr
Strontium
85
At
41
Nb
Niobium
52
Te
(281)
Ds
22 47.867
Ti
Titanium
51
Sb
6 12.011
C
Carbon
Cn
Tl
(278)
Mt
Meitnerium
18 39.948
Ar
Argon
30 65.38
Zn
Zinc
60 144.24
Nd
Neodymium
47 107.87
Ag
Silver
40 91.224
Zr
Zirconium
15 30.974
P
Phosphorus
14.007
N
Nitrogen
86
Rn
82 207.2
Pb
Lead
39
Y
46 106.42
Pd
Palladium
77 192.22
Ir
21
Sc
3 6.94
Li
Lithium
28 58.693
Ni
Nickel
94
34
Se
53
I
78 195.08
Pt
Platinum
49 114.82
In
Indium
5 10.81
B
Boron
44 101.07
Ru
Ruthenium
14
80 200.59
Hg
39.098

IT'S ELEMENTARY

As we have seen, in the nineteenth century atomic theory finally made it firmly back into the mainstream. English chemist John Dalton explained the complexity of matter as a combination of atoms of different elements. Each element's atom had a different weight, from the lightest hydrogen upward. Dalton produced a table of his elements that was wrong in a few details—he thought, for example, that a number of compounds were elements—but it was surprisingly good given that Dalton was working with equipment that was poor even by the standards of the day.

Dalton's table

John Dalton represented each substance he believed to be an element with a symbol and gave each a relative weight compared with hydrogen.

He also speculated on the way that some elements combined to produced compounds, materials where two or more elements are always conjoined in the same proportions. Again, he wasn't entirely right as he assumed incorrectly that these elements would combine in the lowest possible numbers. So, for example, while he knew that water combined hydrogen and oxygen, he assumed a water molecule had just one of each, when we now know it has twice as many hydrogen atoms in the familiar formula H_2O.

Dalton laid out the foundation of modern atomic theory, but was not able to provide the next level of structure. By this time, there was an awareness that some elements bore similarities to each other. For example, the metals sodium and potassium—both isolated by English scientist Humphry Davy in 1807—reacted energetically when they came into contact with water. It seemed likely that there was some kind of underlying structure reflected in the nature of the elements and before long scientists were suggesting how this could be arranged.

JOHN DALTON: A largely self-educated man, Dalton spent most of his working life as a private tutor, but he managed to study meteorology and physics, as well as chemistry.

Two examples along the way were the suggestions of German chemist Johann Döbereiner and English chemist John Newlands. In the 1820s, Döbereiner spotted similarities between several groups of three elements, which he called triads; for example, the metals lithium, sodium, and potassium, or the reactive chemicals chlorine, bromine, and iodine. Unfortunately, it wasn't possible to extend this pattern across all the known elements. The same problem occurred for Newlands who, in the 1860s, followed an example of Isaac Newton.

It's Newton that we can thank (or blame) for the idea that the rainbow contains the seven colors red, orange, yellow, green, blue, indigo, and violet. There is no good reason for coming up with that number of seven colors, and it's thought that Newton probably assumed there was some kind of special natural character to the number, paralleling the seven notes in the musical octave. Similarly, Newlands tried to build up a pattern where the elements had seven different types as weights increased, coming back to be similar again on the eighth element or octave.

VALUING VALENCE

Both Döbereiner and Newlands (along with several others) made the mistake of taking a philosophical approach—trying to impose an arbitrary pattern on nature—rather than the scientific one of looking for a pattern with an open mind and seeing what was in the data. One distinction that was coming to be recognized in the second half of the nineteenth century was that of valence (though it wouldn't get the name until the 1880s).

Valence is a pattern in the way that different elements combine with each other. Rather than being able to be stuck together in any combination, a particular element will typically combine with specific numbers of other atoms. One way to look at valence is how many of a simple element, such as hydrogen that is considered to have a valence of 1, it combines with. So, for example, as we have already seen, oxygen combines with two hydrogen atoms, giving oxygen a valence of 2. Nitrogen combines with 3 and carbon 4. These valences then extend beyond the links with hydrogen. So, for example, carbon dioxide gives us two valence 2 oxygens with one valence 4 carbon.

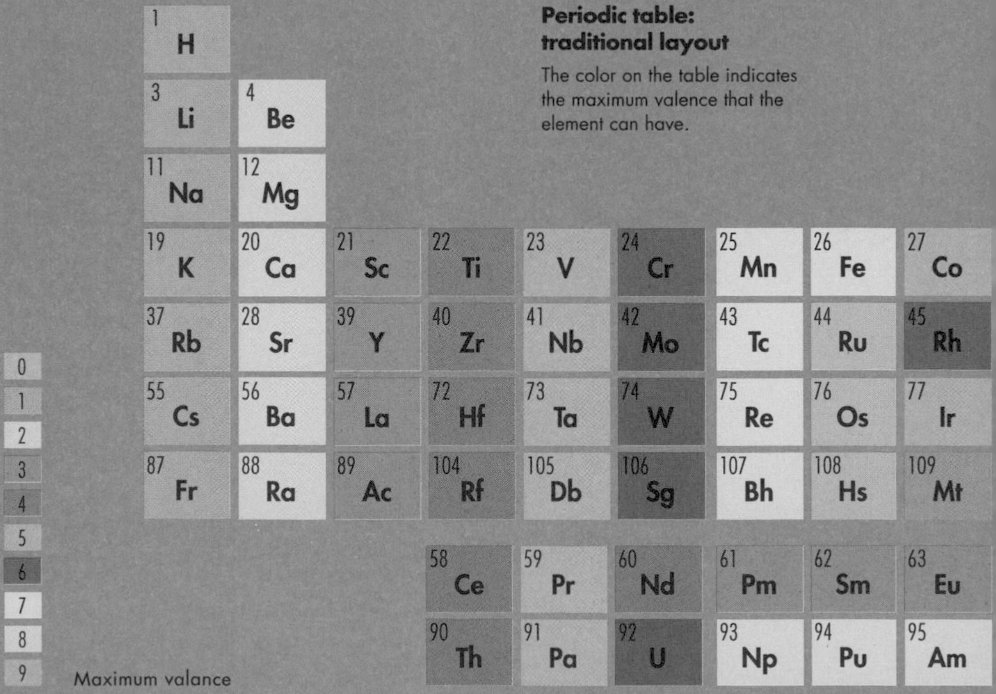

Periodic table: traditional layout

The color on the table indicates the maximum valence that the element can have.

1 H								
3 Li	4 Be							
11 Na	12 Mg							
19 K	20 Ca	21 Sc	22 Ti	23 V	24 Cr	25 Mn	26 Fe	27 Co
37 Rb	28 Sr	39 Y	40 Zr	41 Nb	42 Mo	43 Tc	44 Ru	45 Rh
55 Cs	56 Ba	57 La	72 Hf	73 Ta	74 W	75 Re	76 Os	77 Ir
87 Fr	88 Ra	89 Ac	104 Rf	105 Db	106 Sg	107 Bh	108 Hs	109 Mt
			58 Ce	59 Pr	60 Nd	61 Pm	62 Sm	63 Eu
			90 Th	91 Pa	92 U	93 Np	94 Pu	95 Am

0
1
2
3
4
5
6
7
8
9
Maximum valance

Although things aren't as simple as that description sounds—there is, for example, also carbon monoxide with just one oxygen to each carbon—it provided a natural pattern that could be used to build up a table of relationships better based on reality than Newlands' octaves. German chemist Lothar Meyer, working around the same time as Newlands made a less aesthetically pleasing but more accurate table by simply grouping elements together by their valances.

At the same time, in Russia, the scientist Dimitri Mendeleev was also exploring the structure implied by valence and other characteristics of elements. His first attempt to produce a table was in 1869 and within a couple of years he had developed something that bore more of a resemblance to the modern table (albeit with the rows and columns transposed). Such was the power of the pattern that Mendeleev believed he had revealed the existence of elements that had yet to be discovered.

NOBEL PRIZE FOR CHEMISTRY: In 1906 Mendeleev was proposed for the prize but lost out for political reasons.

Each of the gaps that Mendeleev found was named after the element above it in the table with "eka" (Sanskrit for "one") in front of the name. So, for example, Mendeleev predicted the existence of what we now call germanium, naming it ekasilicon, 17 years before the element was discovered.

								2 He
			5 B	6 C	7 N	8 O	9 F	10 Ne
			13 Al	14 Si	15 P	16 S	17 Cl	18 Ar
28 Ni	29 Cu	30 Zn	31 Ga	32 Ge	33 As	34 Se	35 Br	36 Kr
46 Pd	47 Ag	48 Cd	49 In	50 Sn	51 Sb	52 Te	53 I	54 Xe
78 Pt	79 Au	80 Hg	81 Tl	82 Pb	83 Bi	84 Po	85 At	86 Rn
110 Ds	111 Rg	112 Cn	113 Nh	114 Fl	115 Mc	116 Lv	117 Ts	118 Og
64 Gd	65 Tb	66 Dy	67 Ho	68 Er	69 Tm	70 Yb	71 Lu	
96 Cm	97 Bk	98 Cf	99 Es	100 Fm	101 Md	102 No	103 Lr	

Dark lines on the spectrum show where light of a particular frequency is being absorbed when the light passes through an element in gas form.

WHAT LIES BENEATH

What Mendeleev and his contemporaries had no clue about was the reason the periodic table works at all, or why it has its oddly asymmetric structure. The reason was a pattern that is not conceived by human minds, but that goes to the very fundamentals of nature—the distribution of electrons in an atom.

As we have seen, Rutherford's team discovered the atomic nucleus, making it seem as if the structure of the atom was similar to that of the solar system, with the nucleus being the equivalent of the Sun and the electrons flying around it in orbits like planets. Yet over 100 years ago, this model was dismissed. Its simplicity and likable parallel with a solar system have meant that to this day it remains the most common graphic image for the structure of the atom, but the reality is very different.

We owe our understanding to the development of quantum physics, driven by the work of Danish physicist Niels Bohr, who neatly began this advancement when at Manchester, in England, working for Rutherford. Bohr realized early on that a solar system model would not work, because, unlike planets, electrons are electrically charged. When an electrically charged particle like an electron is accelerated it emits energy in the form of photons. This, for example, is how a radio aerial works—accelerating electrons give off photons in the radio energy range.

Being in an orbit involves constant acceleration. This is not because the orbiting body is getting faster in its orbit, but because it keeps changing direction, and that is itself a form of acceleration. As a result, an electron zipping around a circular or elliptical path would pour out energy and collapse into the nucleus.

Bohr made use of the central idea of quantum physics that energy can't be given off in continuously variable amounts, but rather in quantized packets. He suggested that electrons did not orbit, but had fixed energy levels that they could jump between by emitting a single photon at a time. They could not exist at the levels in between, so would not spiral inward.

This idea was supported by a known phenomenon that particular atoms gave off specific colors when heated, or absorbed specific colors when light passed through them. This was the basis of the science of spectroscopy, which means that we can discover the elements in a star despite never going anywhere near it. In fact, the element helium was discovered in the Sun (named from *helios*, the Greek for the Sun) as a result of spectroscopic analysis. Bohr's theory explained why only certain colors were given off or absorbed, as the color of light corresponds to the energy of photons. So, if only certain energy jumps were allowed, these would correspond to specific colors.

We now think of electrons existing in "shells"—these are different levels that they jump between, each of which can house a number of "orbitals." These are not orbits, but different probability distributions for the location of the electron over time. The available orbitals depend on the number of properties of the electron.

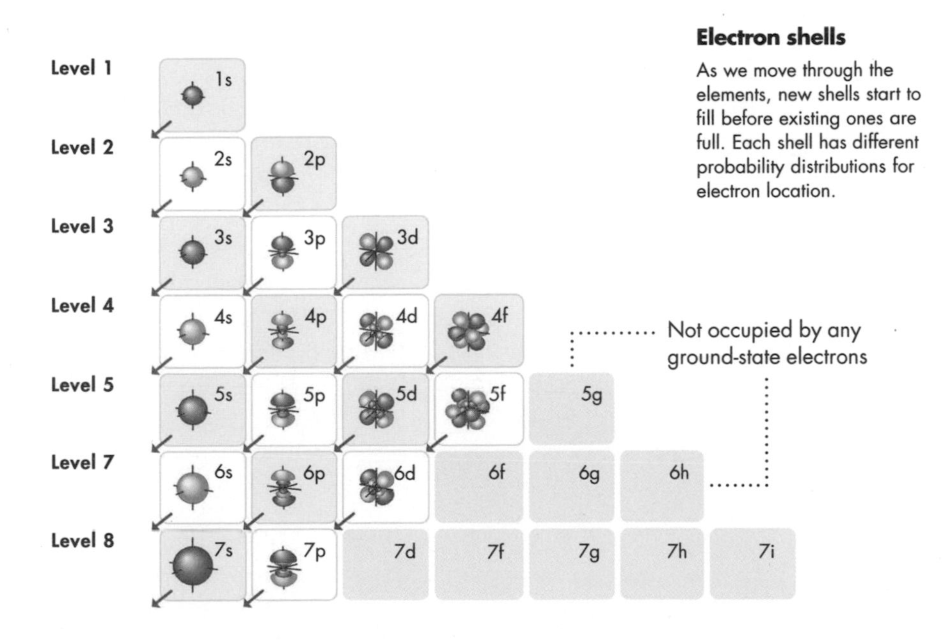

Electron shells

As we move through the elements, new shells start to fill before existing ones are full. Each shell has different probability distributions for electron location.

THE ELECTRON PATTERN

It is the number of electrons in the outermost shell of an atom that gives it its chemical properties; and it is the way that electrons can build up in these shells that is reflected in the rather odd pattern of the periodic table. This is because the way that electrons are distributed as elements become heavier and heavier is not entirely intuitive.

The first shell outside the nucleus is limited to two electrons. This is why the periodic table has those strange bits sticking up at the top for hydrogen, with one electron in that first shell, and helium with a full shell of two electrons. Having a full outer shell makes an element particularly unlikely to react with other elements. This explains why the elements on the right-hand side of the periodic table, the noble gases, are particularly unreactive.

The next shell outward, the second shell, holds a total of eight electrons, making for a fairly straightforward second period (row) of the table containing eight elements, ranging from having one electron in that outer shell (lithium) to eight (neon) in the second of the noble gases. Then the complexity increases. You might think from the rather odd pattern of the periodic table (see pages 100–101) that the third shell also holds eight electrons, because the third period also contains eight elements, but in reality, this shell can hold 18 electrons. However, the stable limit when it's the outer shell is eight electrons, regardless of how many it can actually hold.

As a result of this, what happens is that the third shell fills to eight, and then a new, fourth shell starts to be filled. Once this occurs, what had been the outer shell no longer holds that position, so it can continue to fill up to its total capacity. The result is a staggered filling process. The fourth period has 18 elements in it—getting the fourth shell up to eight and filling the remaining ten spaces in the third shell within it. Similarly, the fifth period has 18 elements by adding ten to the fourth shell and eight to the fifth shell. But even then, the fourth shell isn't full, as it has a capacity of 32 electrons. So, by the time we get to the sixth period there are a total of 32 elements finally filling the fourth shell, partially filling the fifth shell, and getting eight electrons in the outer shell.

The size of the sixth period isn't immediately obvious in the conventional representation of the periodic table as a chunk of

14 elements from cerium to lutetium are extracted as a separate bar at the bottom. These so-called lanthanides—named after the element to the left of the discarded chunk—are partly removed to make the width of the periodic table manageable. This structure also emphasizes the way the outer shell fills up last, meaning that the remaining elements in the period from hafnium to radon correspond to the equivalent columns in the table. These elements are chemically similar because they have the same number of electrons in the outer shell.

OGANESSON: Element 118 is named after the Russian nuclear physicist Yuri Oganessian, whose techniques were used in the production of elements 106 to 118.

The seventh period accommodates 32 elements using an equivalent filling mechanism. All the elements in this period have now been created, although those to the right-hand side all have very short lifetimes before undergoing radioactive decay. In principle, a further period could be added, but the lifetimes would be so ridiculously short that it's hard to see what benefit there would be. Oganesson, the current heaviest element at atomic number 118, has a lifetime of under one millisecond and only six atoms have been created at the time of writing.

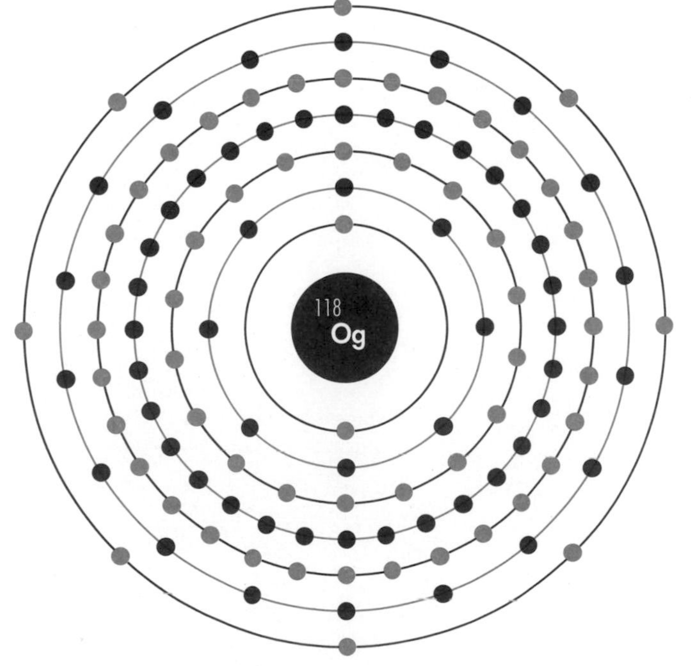

Element 118

The electron structure of the heaviest known element, oganesson, which has a total of 118 electrons. Such ultra-heavy elements are unstable because their nucleus is so big, the short range strong force cannot keep them together.

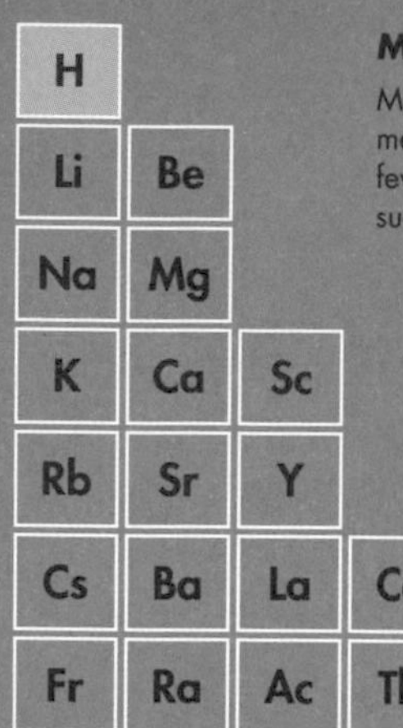

Metals to nonmetals

Most elements are either metals or nonmetals, with a few intermediate "metalloids," such as silicon and germanium.

METALS AND NONMETALS

We have seen how the pattern of electrons on the outside of an atom influences the element's chemical behavior and hence the structure of the periodic table itself. With the table assembled we can begin to see a number of different groupings of elements taking shape.

One that we use regularly, without necessarily thinking about what this grouping really means, is the distinction between metals and nonmetals. In general use, we tend to think of metal as hard but malleable shiny stuff that usually conducts electricity. Astronomers are far more prescriptive: they use the term "metal" for every element that isn't hydrogen and helium. This is because the "base" fusion process in a star is converting hydrogen to helium, reflecting their opening positions in the periodic table. Eventually, the star will go on to fuse heavier elements, and once this happens it is considered to be producing "metals."

In the more conventional sense of the term, ignoring the outlier of hydrogen, everything to the left of a diagonal line that runs more than halfway across the table is a metal, there is a diagonal band of "metalloids" (colored pink in the table above), and then a significantly smaller chunk of nonmetals. Those metalloids are where we find the semiconductors silicon and germanium.

This pattern reflects the way that substances conduct electricity. In conductors, such as metals, one or more electrons are only loosely connected to an atom and can move relatively freely through

														He
									B	C	N	O	F	Ne
									Al	Si	P	S	Cl	Ar
Ti	V	Cr	Mn	Fe	Co	Ni	Cu	Zn	Ga	Ge	As	Se	Br	Kr
Zr	Nb	Nb	Tc	Ru	Rh	Pd	Ag	Cd	In	Sn	Sb	Te	I	Xe
Hf	Ta	Ta	Re	Os	Ir	Pt	Au	Hg	Ti	Pb	Bi	Po	At	Rn
Rf	Db	Sg	Bh	Hs	Mt	Ds	Rg	Cn	Nh	Fl	Mc	Lv	Ts	Og

the crystal lattice of the substance. In semiconductors, it usually takes some extra boost, for example from an incoming photon of light to give an electron enough energy to break free and conduct, while in a nonmetal the electrons don't have the same propensity to break free and conduct.

We have already seen how there are some significant similarities in the columns (known as groups) of the periodic table—it's how it was first assembled. So, for example, we find the likes of lithium, sodium, and potassium, or magnesium and calcium, or silver and gold, or fluorine, chlorine, bromine, and iodine, or the noble gasses in the rightmost column, having distinct similarities. However, it's rare that surface-level similarities continue through an entire group.

> **“EVEN A SCIENTIST CAN'T HELP THINKING OF THE PERIODIC TABLE AS A ZOO OF ONE-OF-A-KIND ANIMALS CONCEIVED BY DR. SEUSS.**
>
> NEIL DEGRASSE TYSON

WHAT'S IN A NAME

Broadly speaking, the names of the elements divide into four sets: traditional ones that have been recognized for longer than the existence of elements themselves; relatively early "scientifically discovered" elements; obscure mid-range elements; and man-made elements than do not occur in nature, which include the most recent entries.

In those golden oldies, we quite often find that the chemical symbol—the one- or two-letter identifier on those periodic table tiles—doesn't necessarily fit with the modern name of the element, as it makes use of classical terminology. So, for example, iron is well known to be Fe from the Latin *ferrum*. Similarly, copper has a slight shift to Cu from the Latin *cuprum*, and others are downright obscure, such as Ag for silver (Latin: *argentum*), Au for gold (Latin: *aurum*), tin Sn (Latin: *stannum*), and lead Pb (Latin: *plumbum*). And when you thought all you needed was a Latin dictionary, mercury takes a leap to Hg for hydrargyrum, which is not real Latin, but a latinized version of the Greek *hydrargyros*.

The Greek goddess Pallas Athene lent her name to the asteroid Pallas and thence to Palladium.

The names for the early elements first isolated by scientists were sometimes derived from substances containing them. So, for example, fluorine,

"IT'S VERY BEAUTIFUL. AND IT'S TRUE.

C. P. SNOW ON THE PERIOD TABLE

not isolated until 1886, was named after fluorite, a mineral containing it in the form calcium fluoride. Fluorite comes from the Latin *fluere*, "to flow," as it was used as flux in smelting iron.

Others from this period derived from the imagination of the scientists involved. The next element down the periodic table from fluorine is chlorine, which was named by Humphry Davy from a Latinized version of the Greek *khloros*, meaning pale green, the color of the gas. Similarly, the malodorous bromine gets its name from the Greek for "stench" and iodine from the Greek for a violet color. Bromine wasn't the only one to have its smelliness highlighted—osmium comes from the Greek for "smell." When not using the impact on the senses (for example in the iridescent element iridium), the source of inspiration could be a little more obscure; Palladium was derived from the name of the asteroid Pallas, itself named after the Greek goddess Pallas Athene.

Later in the nineteenth century, places started to creep into the naming. Perhaps the most difficult to name element was germanium. Its 1886 discoverer, Clemens Winkler, intended to call it neptunium to honor the discovery of the planet Neptune. He then discovered that someone else had already given the name to an element. As it happened, this discovery proved incorrect, so the name neptunium came free again, and would eventually be used to join uranium and plutonium as planet-inspired radioactive elements. Winkler instead went for the newly unified Germany as an inspiration. (Although the country was called Deutschland, the Roman name for part of the region was Germania.)

Other location-based names in the earlier elements include scandium (for Scandinavia), europium for Europe, polonium for Poland, and the magnificent collection of yttrium, terbium, erbium, and ytterbium, all named after the small Swedish village Ytterby, where minerals containing the elements were found. The most entertaining is arguably thulium, named after Thule, which is sometimes pronounced Tooli, although it looks as if it should be Thool, which sounds more suitably dark and mysterious. Originally this was the classical name for a mysterious land, six days sail to the north of Britain, thought by the Greek historian Polybius to be the most northerly part of the world. The element was named in error by Per Teodor Cleve. He not only thought that Thule was the ancient name of Scandinavia, he tried to call the element thullium, but the second "l" was dropped.

In the more recent, synthesized elements that don't exist in nature, the names split into three, named after their place of discovery, their discoverer, or simply a great scientist. Location-based elements include americium, berkelium, californium, lawrencium, and livermorium (after the Lawrence Livermore laboratory), darmstadium (after Darmstadt), and nihonium (from Nihon, the Japanese for Japan).

The discoverers are a more compact bunch. There's seaborgium for Glenn Seaborg, who discovered a remarkable ten synthetic elements and oganesson for Yuri Oganessian who came up with six of these super-heavy elements (see page 105).

Finally, there are those named after other physicists. For example curium (Marie Curie), einsteinium (Albert Einstein), fermium (Enrico Fermi), mendelevium (Dimitri Mendeleev), rutherfordium (Ernest Rutherford), bohrium (Niels Bohr), meitnerium (Lise Meitner), roentgenium (Wilhelm Röntgen), and copernicium (Nicolas Copernicus).

PERIODIC TOPOLOGY

The familiar periodic table based on Mendeleev's layout may be an iconic pattern, readily recognized from its outline, but it isn't the only or necessarily the best way to lay out the elements, and many other structures have been tried. It's a bit like world map projections. When we put the map of the spherical Earth onto a flat sheet (see pages 48–49), there are a number of choices as to how the three-dimensional surface should map onto two-dimensional paper. Similarly, there is not just one way to portray the pattern implied by the elements' electronic structures.

As we've already seen, it is possible to incorporate the extra elements of the lanthanides and the actinides into expanded sixth and seventh periods, giving the table a full 32 columns. But that's only the beginning. Some reformatting takes a more explicit lead from the way the electron shells are built up. So, for example, the so-called "left-step" periodic table adds in elements to correspond to the way the different shells and subshells are filled, producing a more elegant structure that still groups elements by their properties, but makes it explicit exactly where the additional electrons are going as we move up through the table.

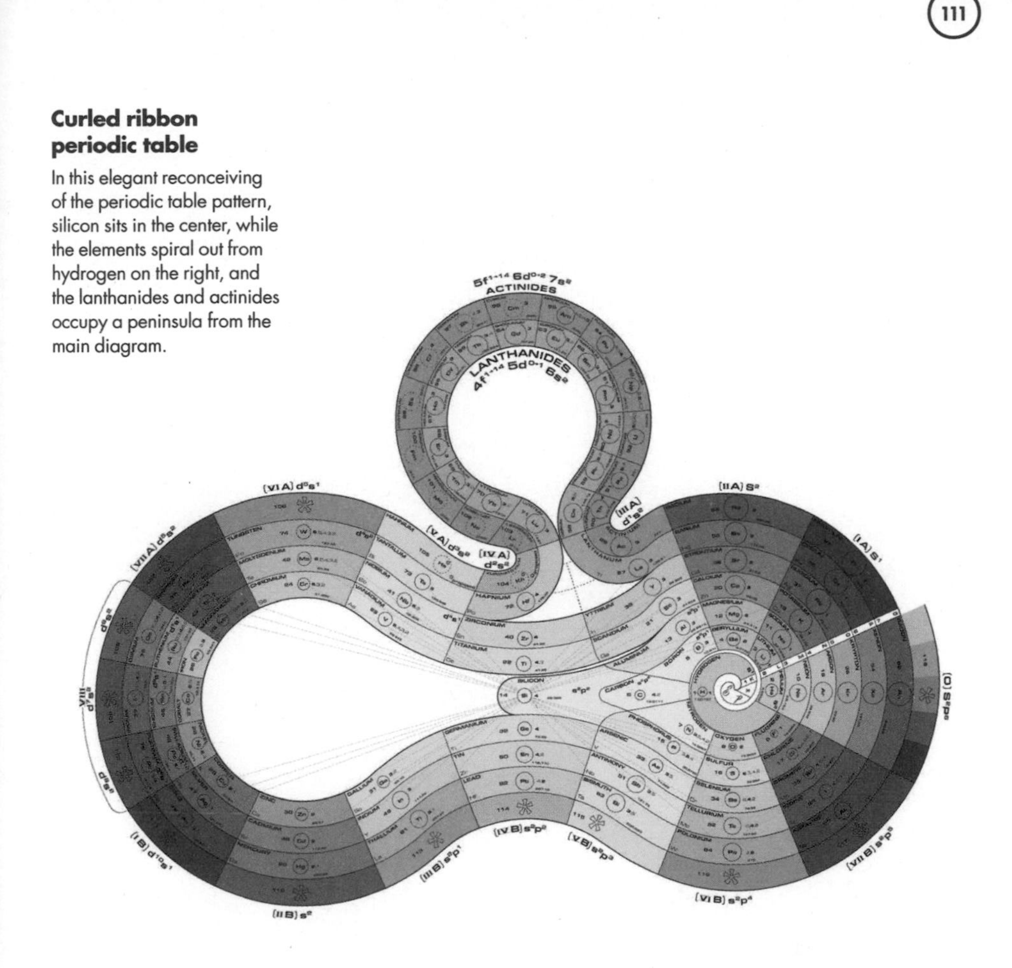

Curled ribbon periodic table

In this elegant reconceiving of the periodic table pattern, silicon sits in the center, while the elements spiral out from hydrogen on the right, and the lanthanides and actinides occupy a peninsula from the main diagram.

Others are more whimsical, adding elements in a spiral form (with extrusions for the lanthanides and actinides), or producing three-dimensional formats, which have included cubes, spheres, pyramids, and more. Some of these are more a triumph of art than providing any real insight into the pattern of the table, but others enable a broader understanding of this essential part of chemistry.

The periodic table represents the way that the chemical elements have gone from being everyday substances, dug up from the earth, to being substances within an underlying pattern, identified by science. Our next pattern has also seen a similar transition. As long as human beings have had consciousness, we are likely to have been aware of the weather, but it is only relatively recently that science has uncovered the complex patterns that lie beneath it.

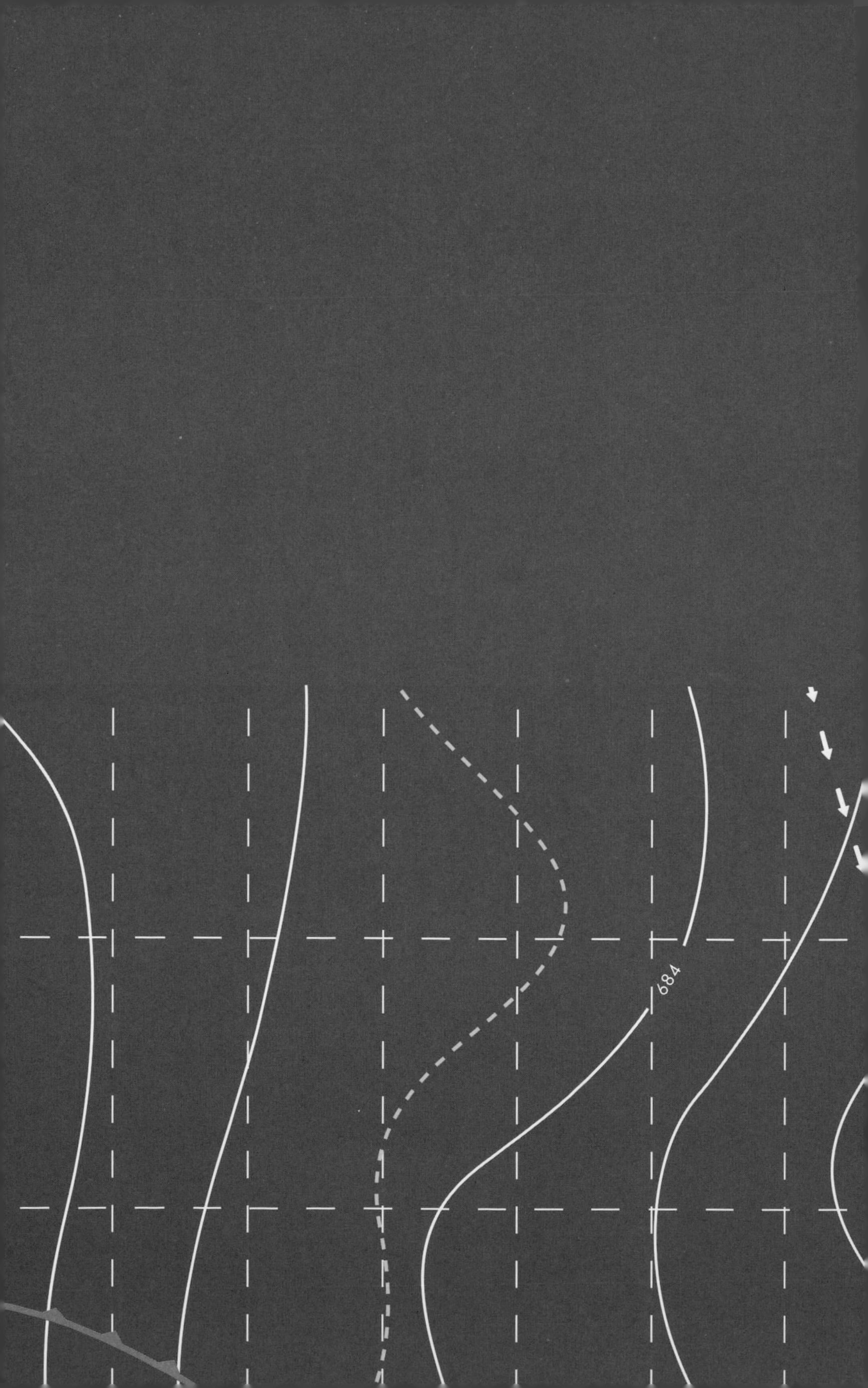
684

6
WEATHER
PATTERNS

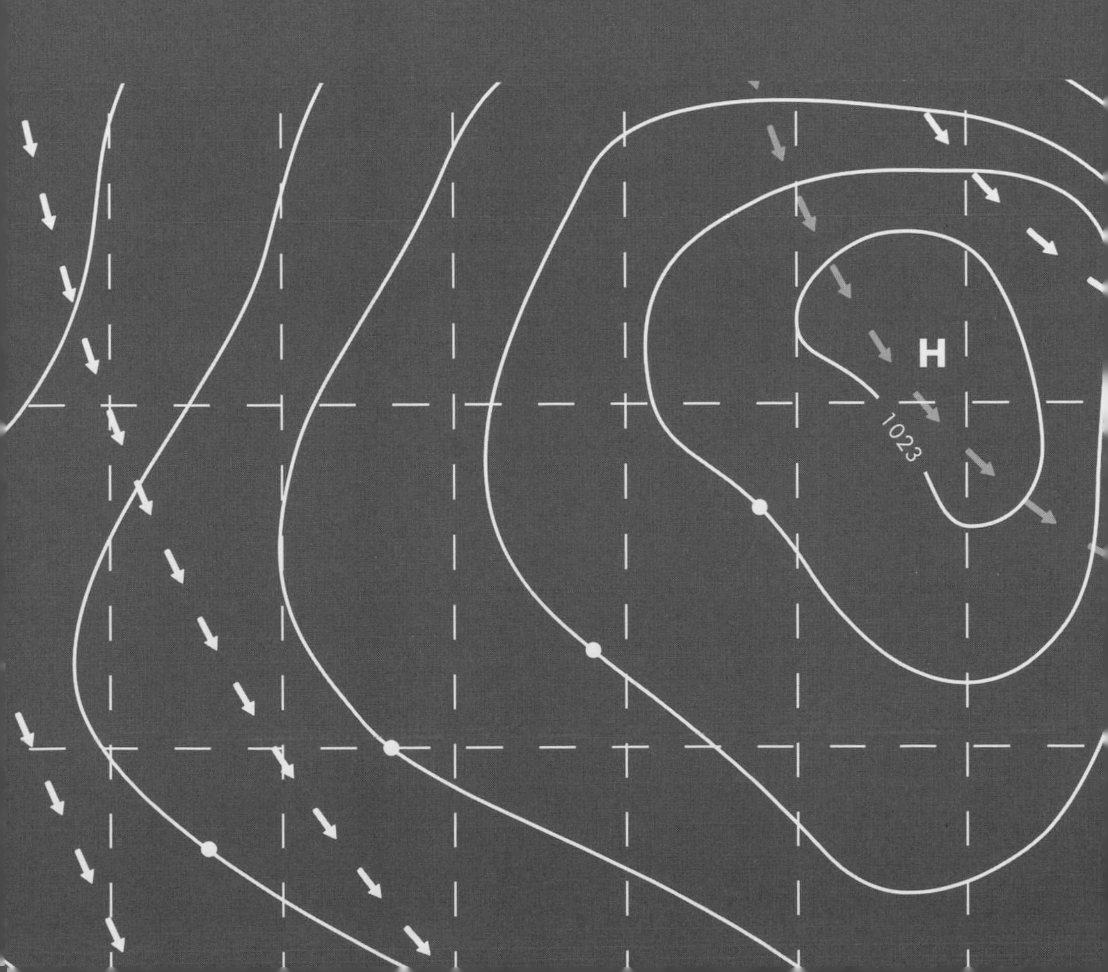
H
1023

Vilhelm **Bjerknes**
1826–1951

UNDERSTANDING WEATHER

Weather forecasting can seem like magic, but underlying the ability to predict what will happen next in such a chaotic system is the awareness of underlying patterns that we can rely on, such as the jet stream in the air and the Gulf Stream in the sea. These large-scale weather patterns have a huge influence on regional climate, while the smaller, faster-changing patterns of high and low pressure bring us the everyday variation that can mean a pleasant day or a life-threatening storm. Although we have had a broad feeling for these patterns for millennia, it was numerical weather prediction, introduced by the Norwegian meteorologist Vilhelm Bjerknes, that made it possible for that understanding to become scientific.

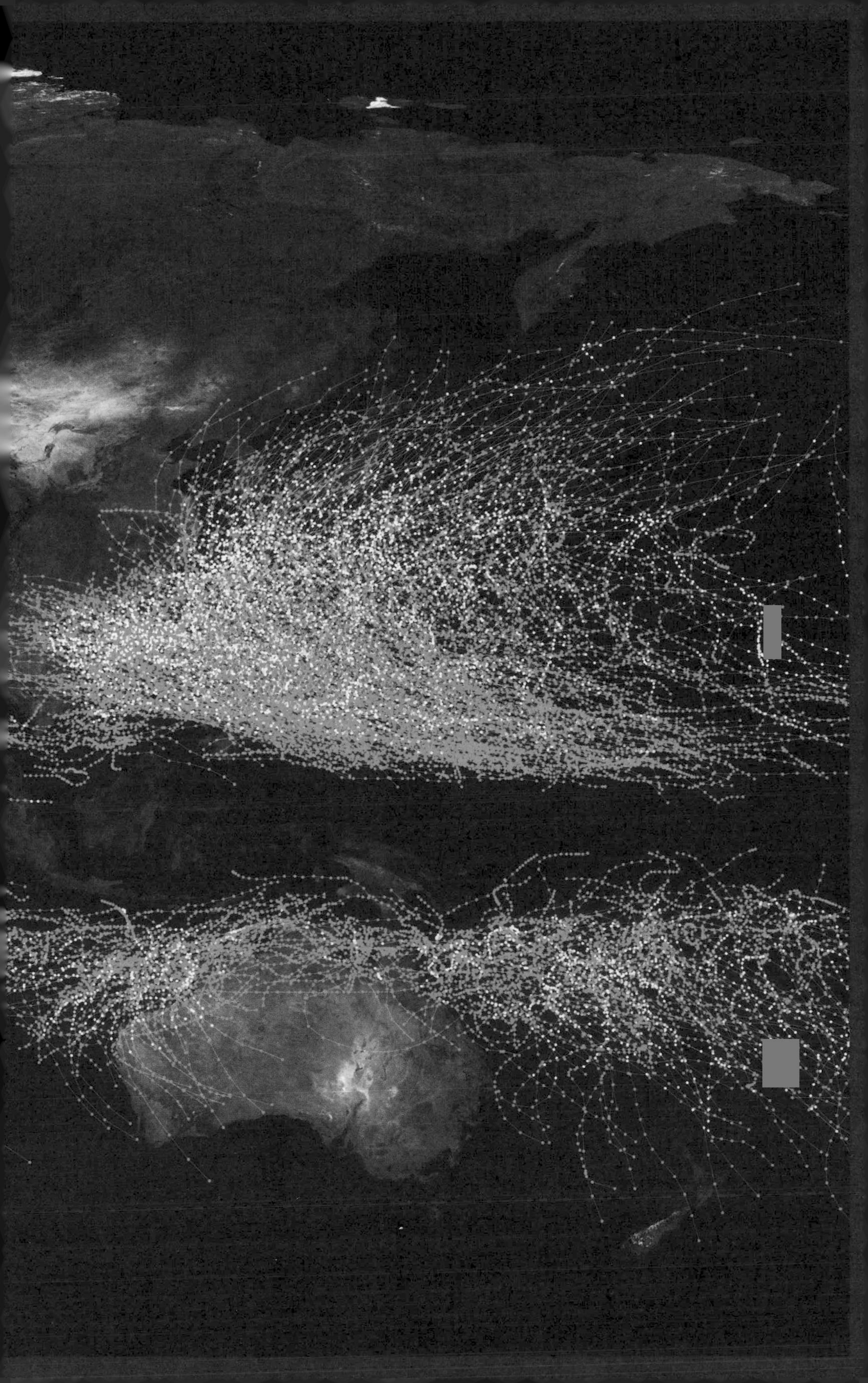

WEATHER MATTERS TO US

All our lives are affected by weather. For those people who work on the land, weather patterns can make the difference between a successful crop and a food shortage, and for anyone experiencing extreme weather conditions, these patterns can make the difference between life and death. This has meant that weather was studied long before the existence of science. We still remember ancient weather traditions, echoing the importance of weather to agricultural societies. Some have a factual basis while others are groundless folklore. Good examples of traditions with a genuine core are the rhyme "red sky at night, shepherd's delight; red sky at morning, shepherd's warning," and the use of seaweed in weather prediction.

The rhyme reflects the importance of air pressure in the pattern of the weather. Red skies occur more often when there is high pressure. If this occurs in the evening the high pressure is often moving in, stabilizing good weather, while in the morning it is typically on the way out, taking the fine weather with it. Seaweed picks up another symptom of the weather patterns: atmospheric humidity. Dry seaweed is good at absorbing moisture from the air, so can become less brittle when humidity rises. Seaweed that regained its flexibility was therefore seen as an indicator of coming rain.

By contrast there are traditions with no relation to reality. These include America's Groundhog Day, when it's alleged that if a groundhog sees its shadow it will head back into its burrow as the winter is due to be extended by six weeks, while if there's no shadow, winter will soon break. This tradition was brought from Germany by the Pennsylvania Dutch (really *Deutsch*, i.e. German); the animal was originally a badger. The performance of the groundhog has been recorded for many years, but fails to give any better prediction than guesswork. Groundhog Day reflects an older tradition that the state of the weather on certain religious festivals shaped the future. The date, February 2, marks the feast of Candlemas, a pivotal point in the early Christian calendar.

"CLEAR MOON, FROST SOON.

TRADITIONAL WEATHER SAYING

Most weather takes place in the lowest of the gas layers of the atmosphere surrounding Earth.

THE IMPORTANCE OF ATMOSPHERE

We take the atmosphere for granted. It may seem little more than a region of gas stuck to the outside of the planet. But, this layer forms a complex system, which is why meteorology—the science of predicting the weather—is anything but easy.

The atmosphere itself is a mix of gasses. Oxygen makes up around 21 percent, while the dominant constituent is relatively inert nitrogen at 78 percent. Include a number of other gasses (notably argon and carbon dioxide), plus water both as vapor and droplets, and a host of particles from soot to bacteria, and you have the constituents for the blanket that surrounds our planet. But this is anything but homogeneous.

The atmosphere is divided into five layers, only one of which has a commonly used name—the stratosphere. This forms the second layer up (hence the use of the word to indicate reaching dizzy heights), the bottom layer, in which we live, is the troposphere. This covers the first 11 miles (18 km) above sea level, reducing to as little as 6 miles (10 km) in mountainous regions. Therefore, the troposphere houses the majority of weather. Because the atmosphere becomes thinner as you move further away from Earth (an inevitable result of the fact that gravity falls off with distance, and it's gravity that holds the atmosphere in place), around 75 percent of all the gas in the atmosphere—and 90 percent of the atmosphere's mass—is within the troposphere. It's within the troposphere that the patterns of weather influence our lives.

UNDER PRESSURE

While folk wisdom may highlight the existence of weather patterns, it takes measurement to pin down what causes them. Many of the measurements necessary for weather forecasting are familiar, but the most important measure—air pressure—seems more mysterious because we don't experience it directly. Air pressure measures the force with which gas molecules hit an area of surface. In a container, the gas molecules shoot all over the place and apply pressure evenly to all sides. But the atmosphere is different. On top of any section of Earth there is a huge column of air and the atmospheric pressure is largely due to the weight of that column.

Of itself, air pressure doesn't mean much to anyone other than a meteorologist. For example, is 1,085 millibars high or low? In fact,

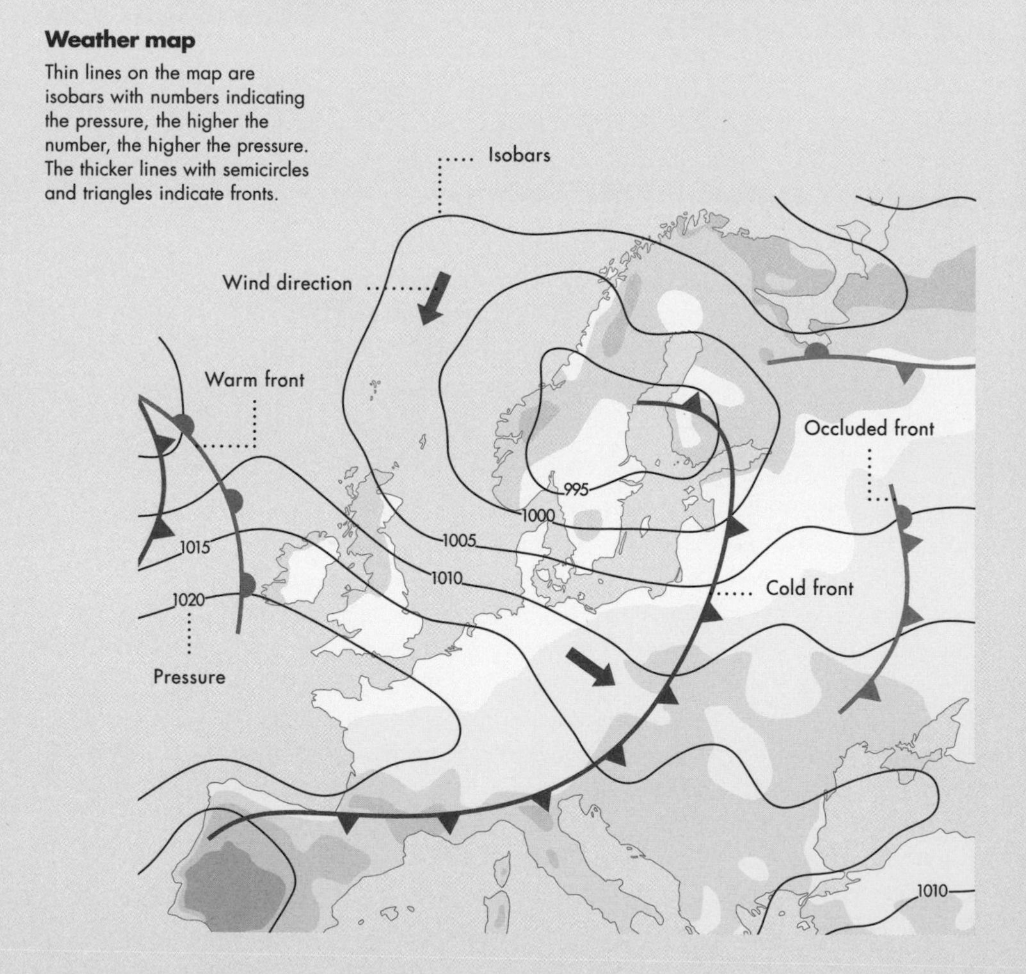

Weather map

Thin lines on the map are isobars with numbers indicating the pressure, the higher the number, the higher the pressure. The thicker lines with semicircles and triangles indicate fronts.

it is the highest recorded air pressure, as a typical pressure is around 1,015 millibars. But the numbers in isolation have limited value: what is important is the relative pressure between two regions, which creates winds and air flow, or the changes in pressure at a location. Pressure levels are shown on weather maps as lines called isobars (see opposite), the equivalent of contours on a traditional map. Isobars pass through places where the pressure is the same. When isobars are close together, it indicates a significant change in pressure over a short distance, usually suggesting a dramatic weather pattern.

MILLIBARS: These are one thousandth of a bar, a unit of pressure that is close to the average pressure at sea level. One millibar is equal to 100 pascals, or 0.0145 pounds per square inch.

Another pattern shown on weather maps is the weather "front." A front represents boundaries between two masses of air with different temperatures and humidity. Such boundaries tend to bring changes in the weather with them, so meteorologists keep a keen eye on their progress.

A warm front forms when warm, wet air from the tropics enters a region of cooler, drier air. As the warm air moves in, because it is less dense, it slides upward over the colder air forming a shallow sloping boundary between the two. At the boundary between the region of warm air and the cooler air it is moving over, the water vapor in the warm air will cool, often resulting in a prolonged band of rainfall. On weather maps, a warm front is usually represented by a series of semi-circles (red, if the map is colored) along the line of the front, which is heading in the direction of the symbols.

A cold front is often found close behind a warm front, as cooler, drier air is pushed into a warmer, wetter section. This undercuts the warm air, typically producing a much more concentrated burst of rainfall than a warm front at the boundary, but then clearing the air, so that after that heavy shower there often follow blue skies. A cold front is represented on a map as a series of triangles (blue, if colored) along the line of the front (see opposite).

A third possibility, quite frequently seen on the maps, is an occluded front. Here, a cold front has caught up with a warm front and undermined it, forcing the warm air further upward and producing a burst of rain and thick cloud. On a map, warm and cold symbols are alternated along the front line of an occluded front.

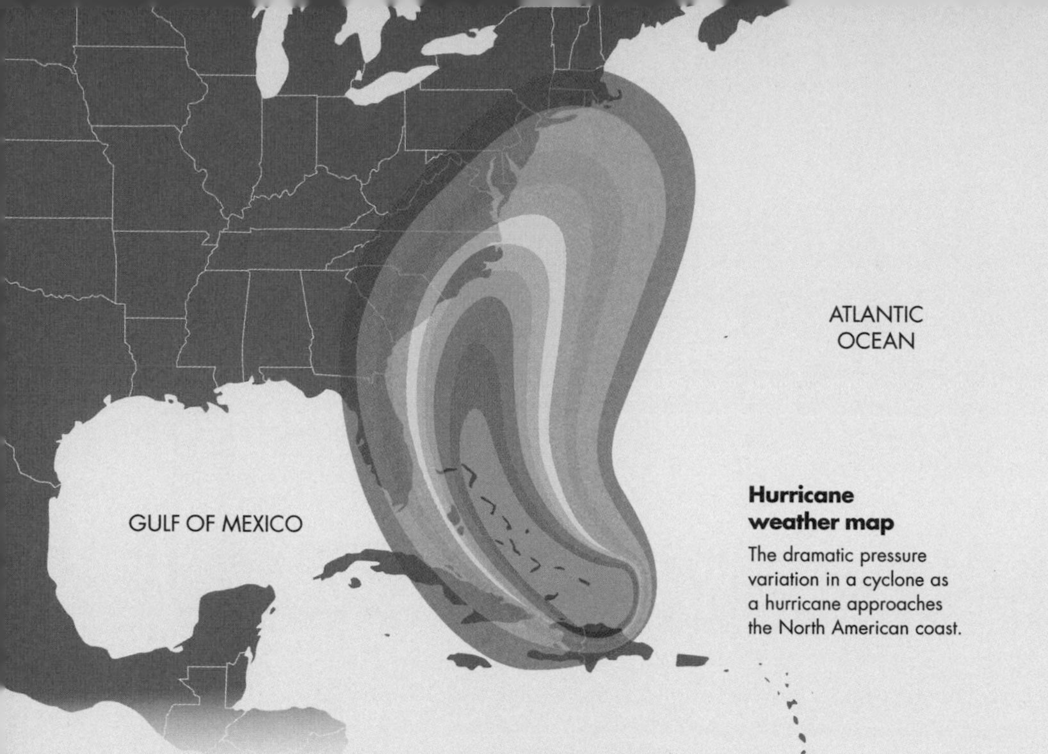

Hurricane weather map

The dramatic pressure variation in a cyclone as a hurricane approaches the North American coast.

Where there is high pressure, associated with good weather, colder air sinks across a region of the Earth's surface (often thousands of miles across) and as it comes closer to the ground, this flowing cold air moves outward, causing a wind that moves away from the center of the region. Because of the Coriolis effect caused by the rotation of Earth (of which more in a moment), this flow of air starts to move in a clockwise spiral in the northern hemisphere and counterclockwise in the southern hemisphere.

Low pressure tends to occur in smaller regions of air over a more compact area. In a region of low pressure, warm air is rising in a column. As it pushes upward it cools, tending to release rain (hence the association of low pressure with bad weather). As the air moves into the column at low levels, it spirals round in the opposite direction to a high-pressure region; in low-pressure regions, air moves counterclockwise in the northern hemisphere and clockwise in the south. Because of these spirals, a low-pressure region is sometimes called a cyclone and a high-pressure region an anticyclone.

The Coriolis force reflects the fact that the air is on a spinning Earth. If Earth were stationary, there would be no rotation of columns of low or high pressure, but because of the turning of the planet, the effect (from our viewpoint) is to see the columns of air being twisted in the opposite direction to the way Earth turns.

SEASONAL PATTERNS

As the atmosphere was studied more, it became clear that some large-scale patterns impact weather around the world. The first to be recognized is such a part of our lives that we tend not to even think of it as a weather pattern: the seasons. We know that Earth travels around the Sun in a near-circular ellipse, but it is tilted. Earth's axis is at an angle of around 23.5° to the vertical, compared with the plane in which it moves around the Sun. This tilt means that for a portion of the year the northern part of Earth is pointing toward the Sun and for the rest of the year the southern part of Earth is oriented more in the sunward direction.

This small difference in orientation produces the seasons. For half the year the Sun's light and energy hits the northern hemisphere more directly, and for the other half it favors the south. This is the reason that December, for example, is midwinter in Europe and North America but midsummer in Australia and South America. In the northern summer, when the northern part of Earth is tilted toward the Sun, the sunlight has slightly less far to travel and has less air to get through before it hits the ground, when compared with the northern winter.

This small tilt is enough to produce most of the differences in weather between the height of summer and the depths of winter. It is most obvious at the poles, which only receive direct sunlight for half the year. So, for instance, the North Pole only sees the Sun between March 21 and September 23. For anyone stationed there, the Sun is below the horizon for the rest of the year in a six-month stretch of twilight and night.

The midnight Sun occurs near one of the poles during the summer, when the Sun is visible all night.

The Gulf Stream makes the UK significantly warmer than many other locations at the same latitude, enabling subtropical plants to grow further north than their usual habitats.

THE BIG PICTURE IN THE AIR AND SEA

Other major patterns are less immediately obvious, as they are invisible fluid movements, but have still been with us for centuries. The earliest to be noticed, if not understood, involved weather on Jupiter. The planet's most outstanding feature is the great red spot, a red patch in its atmosphere bigger than Earth. The red spot has been there for hundreds of years. As we will see later on, the mathematical concept of chaos has a major influence on our modern understanding of weather patterns, and one feature of a chaotic system is that long-lasting islands of calm, representing stable flows of fluid, can spontaneously emerge. The great red spot is such a feature of Jupiter's weather, while Earth's most familiar patterns of this kind are jet streams.

A jet stream is a fast-moving "river" of air created when an area of warm air meets a region of cold air high in the atmosphere, causing a significant pressure difference. There are four major jet streams, running between 6 and 7.5 miles (10 and 12 km) above Earth's surface, distorted out of their natural straight-line flow by the Coriolis effect. Jet streams were first noticed by the Japanese meteorologist Wasaburo Ooishi, who spotted their influence on weather balloons. But the practical impact of these high-speed winds was only discovered in the Second World War when US bombers struggled to lock onto their targets when entering the jet stream over Tokyo. Although flying at 400 miles per hour (650 km/h) with respect to the air, the speed of the wind meant that they achieved 550 miles per hour (880 km/h) over the ground. Airliners make use of a similar effect to cross the Atlantic faster toward Europe.

An equivalent of a jet stream in the water also produces long-lasting weather patterns. The best-known is the Gulf Stream, part of the North Atlantic conveyor. That "conveyor" reflects the system's similarity to a conveyor belt. Winds blowing over the North Atlantic cool the already frigid water, which sinks and flows at a low level toward the equator. At the same time, water on the surface of the sea is being warmed by the Sun in the Gulf of Mexico; this warm water moves north to compensate for the cold water flowing back far below it.

THE GULF STREAM: A warm ocean current that begins in the Gulf of Mexico and flows around Florida and up the US coast before crossing the Atlantic to Europe, to warm some of its northern shores.

The warm current of the Gulf Stream means that northwest Europe is 16°F (9°C) warmer than it otherwise would be, because the climate in this region should be more like Siberia. This process, known technically as thermohaline circulation, transports large amounts of heat from the tropics to northern latitudes. It was the collapse of the North Atlantic conveyor that was portrayed so dramatically in the movie *The Day After Tomorrow*. The scene as portrayed was not realistic, because everything happened much too quickly, but the underlying concept is not fictional.

There is some evidence that climate change can slow down the conveyor. This is because more fresh water is coming into the oceans from melting ice sheets. This addition decreases the density of the cold water that should be diving down and heading south, as fresh water is less dense than saltwater. The result is to reduce the driving force of the conveyor. Unlike the event in the movie, this process is slow, taking perhaps 100 years to reduce the strength of the Gulf Stream by around 25 percent; a change that will be more than balanced out in its impact by global warming.

Jet streams and the Gulf Stream

The air currents that flow in Earth's atmosphere are called jet streams, while the ocean current is the Gulf Stream. They are both long-term, large-scale systems that influence weather patters over great distances.

WEATHER BOYS AND GIRLS

One other large-scale pattern is El Niño ("the boy" in Spanish). This is a pressure seesaw (teeter-totter), where two climate-connected zones experience a regular pattern where pressure rises in one area as it falls in the other. The El Niño–Southern Oscillation is a system crossing the Pacific Ocean, linking high pressure experienced in the Southeast Pacific with low pressure around Indonesia. The result is a change in the trade winds blowing across the ocean and a shift in the location of warm water, which combine to have potentially dramatic effects on local weather conditions.

As El Niño oscillates, the pressure difference between the South American and Indonesian side of the system varies. As the high pressure weakens, the easterly trade winds across the Pacific drop off and can even reverse. Warm water spreads eastward from the Indonesian side, heating up the waters on the western coast of South America. This makes it more likely that there will be heavy rain and flooding along the South American coast. The torrential rainfall can cause mudslides that destroy whole villages.

On the Indonesian side, the result is drier than usual weather, risking drought around Australasia, especially in parts of Australia. The prolonged dry spell arising from El Niño also increases the danger of destructive wildfires breaking out. These can be so dramatic that the wildfires in Indonesia during the El Niño events of 1997, 2006, and 2019 pumped more carbon dioxide into the atmosphere than is normally produced by the entire planet in a year.

Occasionally, the oscillation tips the other way, resulting in cooler water spreading toward Indonesia. Although this is only the other extreme of the oscillation, it is given the name La Niña ("the girl"). When La Niña occurs, it reverses the typical effects, bringing heavy rain to Indonesia and Australia, and dry weather to the South American coast. This massive weather pattern does not just influence the countries directly in its path. The extra warm air produced on the eastern end of the system during El Niño rises up and shifts the jet stream flowing from Japan. This brings drier, warmer weather to the northwest side of North America. The shifts in weather patterns can also lead to an increase in rainfall over East Africa, causing flooding and destroying crops, while West Africa suffers from reduced rainfall and potential drought.

El Niño

During El Niño conditions, a change in distribution of warm water and pressure across the Pacific results in imbalanced weather outcomes.

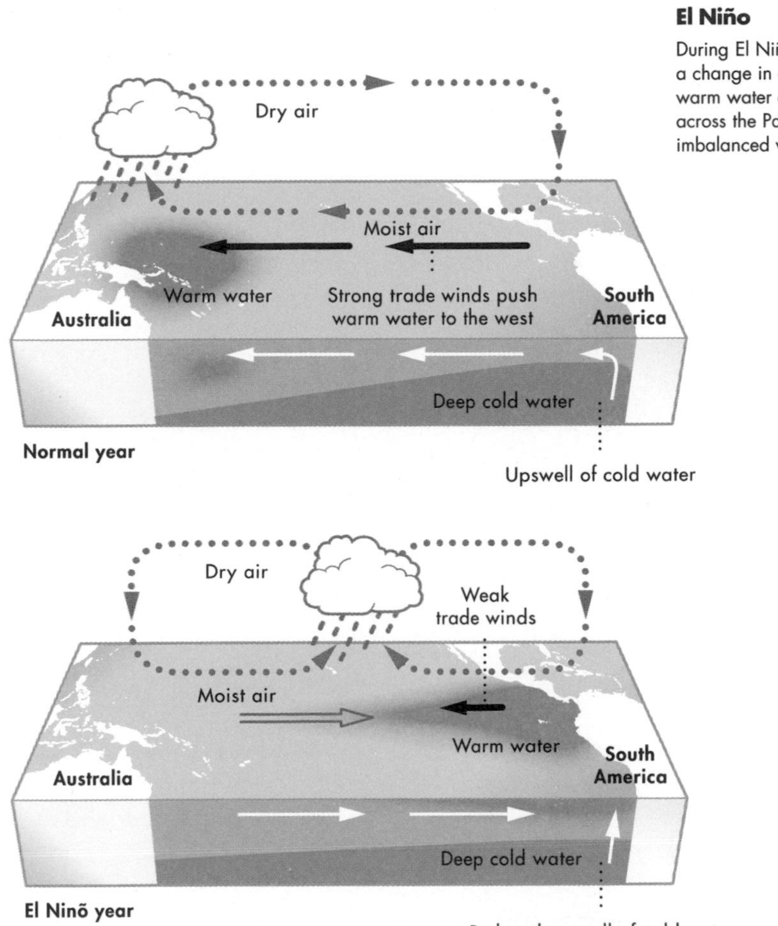

There is an equivalent, weaker, pattern across the North Atlantic, known as the North Atlantic Oscillation. Its El Niño equivalent is the "negative phase" of the oscillation, when there is a weak pressure gradient, while the more frequent positive phase is like La Niña. This produces Northern European winters with relatively wet, mild weather in the North West, leaving the Mediterranean relatively dry. When the oscillation is in negative phase, Northern Europe is hit by much colder winters, while more southerly parts are wetter than usual.

MAPPING THE WEATHER

When it comes to presenting patterns in the weather to the public, the archetypal means is the weather map, which became feasible with the development of the electric telegraph, enabling weather observations across a region to be pulled together to provide an overall picture. *The Times* newspaper in London carried the first weather map on April 1, 1875, showing the weather around the UK on the previous day.

Weather maps are now generated by computers from data gathered by weather satellites, ground stations, and weather balloons. Cheap and easily deployed, balloons are still widely used, with thousands employed every day to monitor weather developments. Aircraft also provide airborne data: many commercial flights carry automated weather stations that relay data to the ground, and flight crew often provide visual observations, typically every ten degrees of longitude on the plane's journey, to help build a picture of the prevailing weather.

Important though these mechanisms are, weather satellites rule the roost. It was less than three years after Sputnik, the original man-made satellite, went into orbit that it was followed by TIROS I, the first of many weather satellites to span the globe. Since TIROS launched in April 1960, sending back crude black and white TV pictures of cloud cover, Earth has been ringed with advanced space

technology, looking down on our weather systems. As well as video, satellites capture infrared images to detect heat emissions from both ground and air. Some satellites add lidar, using lasers in a similar fashion to radar, sending out pulses and watching for reflected photons that are used to detect particles in the air that can reduce light penetration and allow an area below to heat up.

Many of these satellites are geostationary, orbiting 22,000 miles (36,000 km) above Earth's surface. At any particular altitude there is a specific speed required to keep a satellite in orbit. In effect, a satellite falls toward Earth under the pull of gravity, but at the same time it moves sideways at the right speed to keep missing. At just the right speed, the two motions cancel out and the satellite stays in orbit. At 22,000 miles up, the velocity needed to stay in orbit is the same as the rotational speed of Earth, so satellites at this altitude can monitor changing weather conditions across the same section of the surface.

Although weather satellites contribute much of the data on modern weather maps, other information comes from a ground-based equivalent—weather radar. This provides a more localized view than a satellite, producing a picture of the intensity of rainfall for approximately 150 miles (250 km) around the site of the radar source. The radar beams microwaves toward the clouds, and the scattered radiation is picked up back at the source, indicating both location and the strength of rainfall patterns.

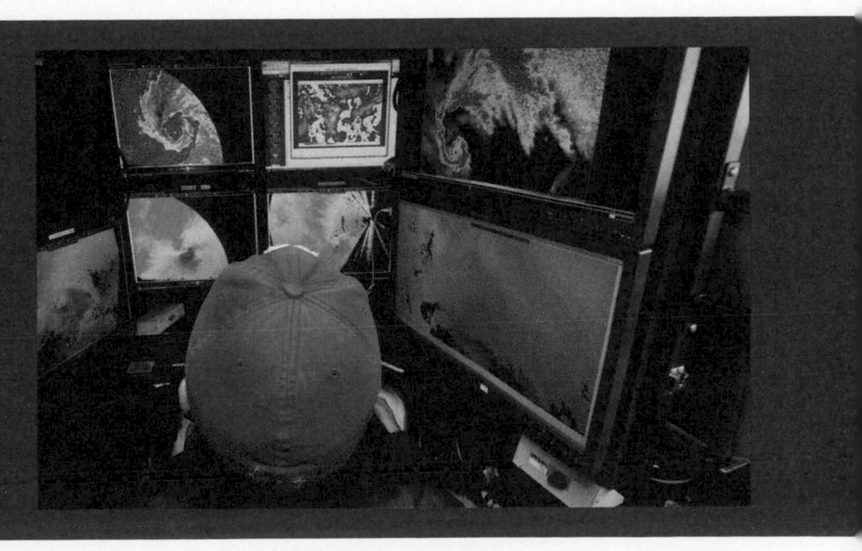

Left: Hurricane Florence seen from the International Space Station.

Right: Computerized controls in a mobile weather radar station.

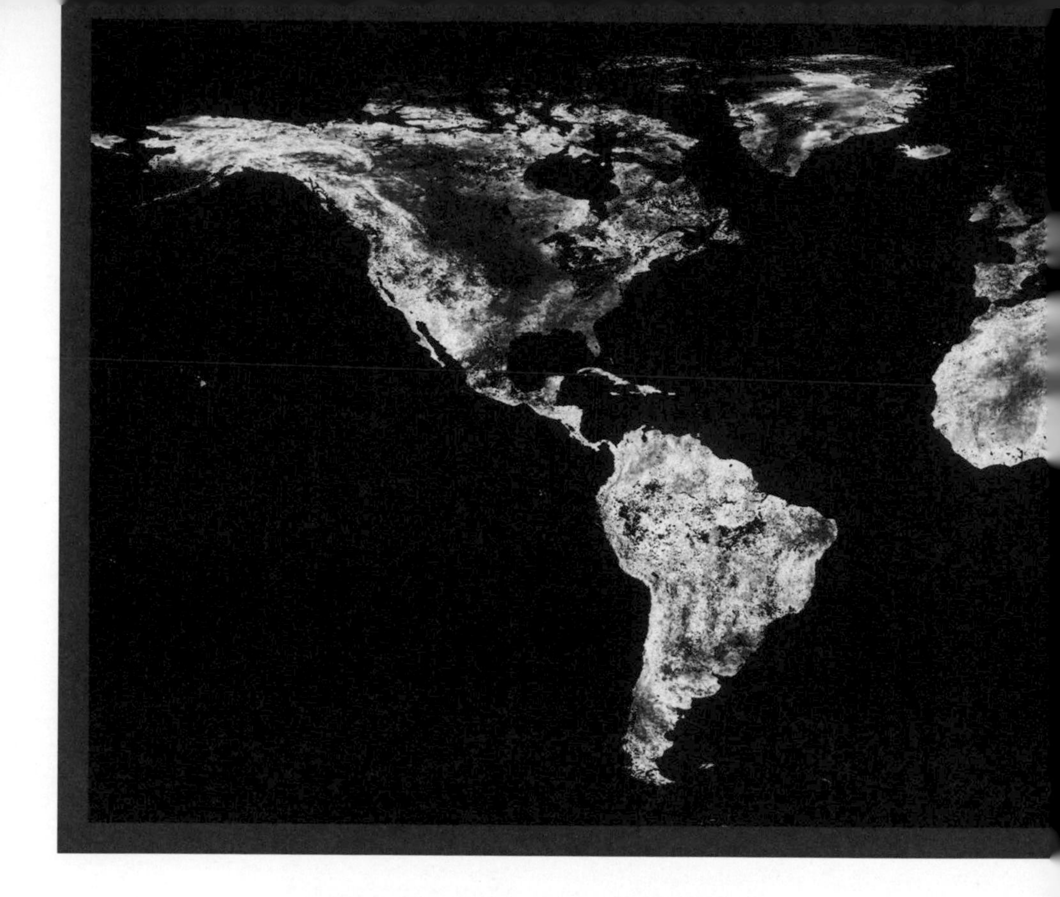

MATHEMATICAL PATTERNS

Twenty-first century weather forecasting involves throwing vast amounts of data at huge supercomputers that undertake billions of calculations a second. These computers simulate the Earth's climate systems, predicting how those systems will evolve over the next few hours or days. Taking a mathematical approach to modeling weather patterns was first attempted by the British mathematician Lewis Fry Richardson during the First World War, but has only become practical with computer support.

Modern computer weather models break up the globe into a set of distorted rectangular cells (the distortion is due to Earth's curvature), each of which is envisaged as having a series of imaginary boxes above it. It's the three-dimensional equivalent of dividing up a flat surface with a grid. The model then works out from current conditions how weather patterns will change from box to box. The smaller and more numerous the boxes, the more accurate the prediction (at least over a short period of time), which is why modern meteorology demands the biggest, most powerful supercomputers.

Worldwide weather

Modern forecasting relies on the ability to pull together weather data from across the world. This weather map shows global temperatures against the average during a particularly cold period.

A huge change happened in the way computerized weather forecasting was undertaken toward the end of the twentieth century, which would allow for significantly greater accuracy in forecasts covering 24 hours to five days. Earlier computer forecasts had worked on a single picture of how things were going to be. Forecasters ran the model, and the output was produced. Unfortunately, weather systems are so dependent on small changes in initial conditions that any particular forecast was likely to be wrong.

To see why it is so difficult to forecast weather patterns accurately you need to take a look at the basic physics underlying it. Back in the seventeenth century, Newton introduced a new view of reality, sometimes called the clockwork universe. The idea was that, given enough data, we should be able to accurately understand how the pattern evolves, perfectly predicting the unfolding weather. However, more recently it has been shown that this will never be feasible.

Weather forecasting will always have a problem, predicted by chaos theory. This is a mathematical field developed in the second half of the twentieth century that has at its heart the idea of a chaotic system. In such a system, a small change in the way factors

The ability to forecast weather more accurately gives greater opportunity for warnings of the impact of dramatic weather patterns, such as those caused by El Niño.

are at the start, results in a very large change after the system has evolved through time. This is often typified as the "butterfly effect"—a small difference like a butterfly flapping its wings can have a much wider impact on the overall pattern. It's no coincidence that chaos theory crops up in predicting the weather. The man behind this theory, American Edward Lorenz, was a mathematician and meteorologist, and it was studying the way small changes in the distribution of temperatures, pressures, and winds could make a huge difference to a weather forecast that brought him to devise this concept.

By allowing for this, a combination of excellent satellite observation and modern forecasting techniques means that forecasts can usually be fairly accurate over 24 hours and make a reasonable attempt at accuracy over three to five days, but anything more than this becomes little more than educated guesswork. Long-range forecasts, no matter how much computing power is thrown at them, remain poor. Most of us have encountered promises of a great summer, only to find that the weather is a flop.

In part, this failure is because a long-range forecast can only give the broad picture, and local weather can be very different from a national average. But it is also impossible to be certain when taking a chaotic system like the weather so far into the future. Even so, the quality of long-range forecasting has improved in the last 20 years, in part because we now have a much better understanding of large scale, long-lasting weather structures such as El Niño and the Atlantic conveyor.

COMPUTERS AND MODELS FIGHT BACK

PROBABILISTIC WEATHER: A forecast predicting a "40 percent chance of rain" is sometimes misinterpreted as rain over 40 percent of the area, or for 40 percent of the time. In reality it means that 40 percent of the model runs predicted rain.

Now, with vastly more computer power, meteorologists run models many times, each with subtly different changes reflecting the uncertainties in the data and how the weather will evolve. The European Centre for Medium-Range Weather Forecasting, for example, which supplies ensemble forecasts around the world, typically runs 50 forecasts per day, each varying slightly in its parameters. The outcomes are grouped together by those with similar results to get a feel for the most likely forecast.

This ensemble approach means that it is possible to predict the probability of different weather events occurring and it is why weather forecasts now show probabilities. A quiet revolution has occurred. Just 30 years ago forecasts were more often wrong than right. Now, short-term forecasts are much more reliable. We may still moan when the weather forecaster gets it wrong, but we have reason for complaint far less often than was the case before ensemble forecasts.

Some mathematicians argue that the impact of chaos on forecasting has been overplayed, as most improvements in forecasts since the 1980s have resulted from having far more detailed mathematical models and computers with extra power to run them on. But there is no doubt that underlying the patterns of weather is the chaotic nature of the weather system.

The weather is a real-world pattern with a huge influence over our lives, but the next example is an imaginary pattern that has, nonetheless, been behind much of the development of modern civilization: the number line.

"CHAOS IS LAWLESS BEHAVIOR GOVERNED ENTIRELY BY LAW. IAN STEWART

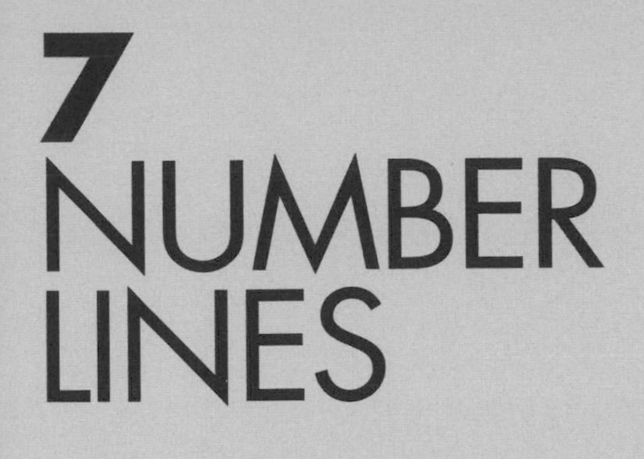

7 NUMBER LINES

Georg **Cantor**
1845–1918

MATHEMATICAL PATTERNS

A number line—a sequence of numbers following a simple rule, such as adding 1 to the previous number—is a remarkably powerful pattern, despite its apparent simplicity. It lies at the heart of set theory, of arithmetic, and even of the mathematics of infinity. In a more sophisticated form it was the basis of the powerful precomputing scientific and engineering tool, the slide rule. Taking the number line up to multiple dimensions brings in complex numbers, which have proved invaluable in mathematical analysis, while going beyond the integers and fractions can bring us to number sequences that lie behind patterns in geometry and nature. Number lines provide the link between the essence of numbers and the ability to take on the patterns of the world with mathematics.

AN INFINITE RULER

The ruler is one of the first pieces of scientific equipment based on a pattern that we encounter. It is marked 1, 2, 3, and so on, at regular intervals. This is, in effect, a small segment of a powerful abstract mathematical concept—an infinitely long pattern called the number line. Imagine a line stretching off into space from your current location. Label your location 0 and then continue with the integers (whole numbers) 1, 2, 3, and so on, marked off along the line. This is the basis for the aspect of mathematics that we make most use of—arithmetic.

SETS: A set is a powerful tool in mathematics. It is a collection of items, which could be real (apples) or mathematical (numbers).

We can see this by thinking of the simplest arithmetical operations: addition and subtraction. Imagine you want to add 9 and 5. What does this mean? In the real world, you might have 9 apples. A friend gives you 5 more. Addition is the action of combining those two sets of apples. (Remember that word "set" as we will come back to it.) If you now count up your apples there are 14. But the power of mathematics is taking a step back from reality to provide a more general pattern that can be applied to any similar task.

"BY THE ANCIENTS, ARITHMETIC WAS STUDIED THROUGH GEOMETRY. IF A NUMBER WAS REGARDED AS SIMPLE, IT WAS A LINE.

LEWIS CAMPBELL

Number line arithmetic

Addition on the number line is simply a matter of moving the required number of spaces to the right.

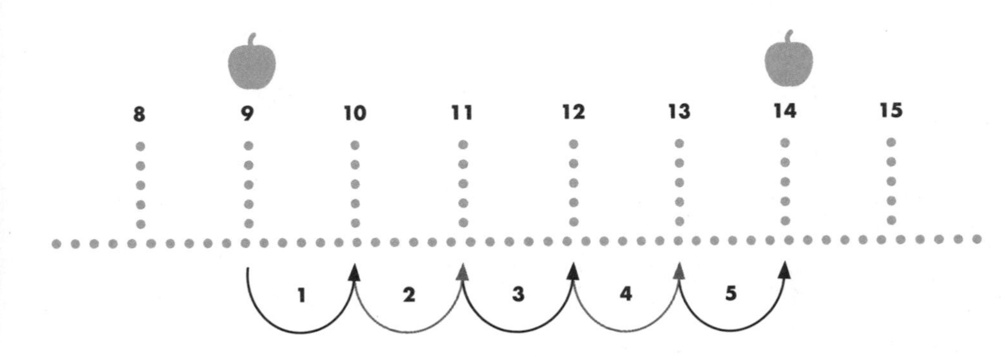

The pattern of the number line provides the mechanism for arithmetic. If you start at 9 on the line and move 5 markings to the right, you will arrive at 14. All that was required to undertake that addition was to take 5 steps along the line (see above). It's all about a shift in the pattern. Similarly, subtraction reverses that process. If someone steals 10 of your apples, you have 4 left. On the number line you start at 14 and move 10 markings to the left, reaching 4. However, going left provides us with a whole new concept. Moving to the right—adding—the number line is endless. If there were a biggest number called TheEnd, you can always move one further space to TheEnd + 1. However, moving a space to the left of 1, you hit 0, where the original line began.

Here, the pattern of the number line goes beyond reality with valuable results. If you had 1 apple and gave it away, leaving you with none (0), it is meaningless to ask what the result of giving away another apple is. But in the mathematical world, it is more fruitful to imagine the line carries on forever to the left, just as it does to the right. Again, we have markings for each whole number. But to indicate we're left of zero, we put a minus (−) sign in front of each. So, we have −1, −2, −3, and so on. We can now move left and right along the whole number line using addition and subtraction.

An abstract sculpture in the form of the leminscate creates a feeling of infinity.

GOING ALL THE WAY

The pattern of the number line never ends. Yet, most children pick up the idea that there is an end. It's not unusual for a child to recite a whole list of increasingly larger numbers before proudly stating "Infinity!" This is, indeed, the limit of the number line, but how is it possible to reconcile the line having a limit with there being no biggest number?

In reality, infinity is not a number at all. It is a target, rather than a place on the repeating pattern of the line. At school, you may have represented infinity using the squashed figure 8 on its side ∞, called the lemniscate. This is a particular type of infinity, identified by the Ancient Greek philosopher, Aristotle. He came up with a neat illustration of its nature, using the Olympic Games.

Aristotle asked "Are the Olympic Games real?" Of course they are. "Then," Aristotle asked, "Show me these Olympic Games." And unless we happen to be at a particular time and place, we can't. There is a pattern to the existence of the Olympic Games, but at our location in that pattern we can't see them. Aristotle described infinity as a *potential* thing. It's not possible to show it, but it has the potential to be, and we can use it once we introduce another feature to the pattern of the number line—fractions.

Rather than only divide our number line by the pattern of the integers, positive and negative, we can also divide each section of the line into smaller parts. And Aristotle's potential infinity then enables us to do something surprising.

Imagine taking one step along the number line to the right, from 0 to 1. Then take half a step from 1 to 1½. Then half that size step, from 1½ to 1¾. And so on, forever. Where do we end up as we head toward potential infinity? It might seem natural that taking an infinite set of steps would reach the end of the line. But at the rate that each step gets smaller and smaller, we will never get past 2. The limit of adding together the numbers 1 + ½ + ¼ . . . is 2. In practice, we can't quite reach 2, but that potentially infinite set of numbers represented by ∞ would take us to 2.

This is a remarkable pattern, which is related to the pattern underlying much of the mathematics of calculus in a process called integration, used to calculate the areas of shapes or swept out by curves. For example, to calculate the area of a circle we can imagine dividing up the circle into a pattern of segments. Each segment is approximately a triangle, for which it is easy to work out the area. To get the exact result, we imagine having more and more triangular segments, which are thinner and thinner. In the limit of an infinite set of infinitely thin segments we reach the area of the circle.

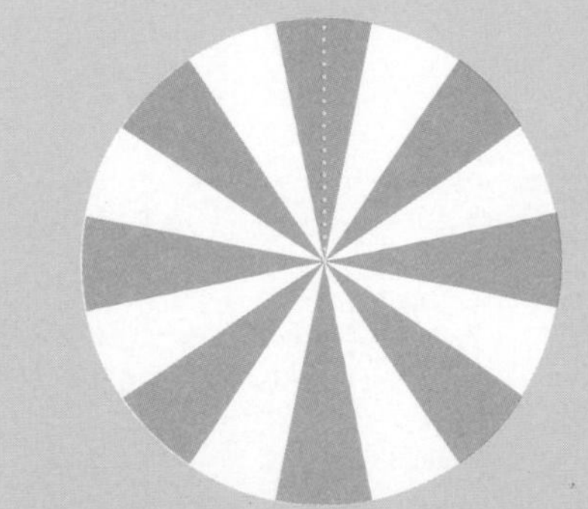

Areas from infinity

Approximating the area of a circle by dividing it into segments. Taking more, thinner segments means each segment becomes closer and closer to a triangle.

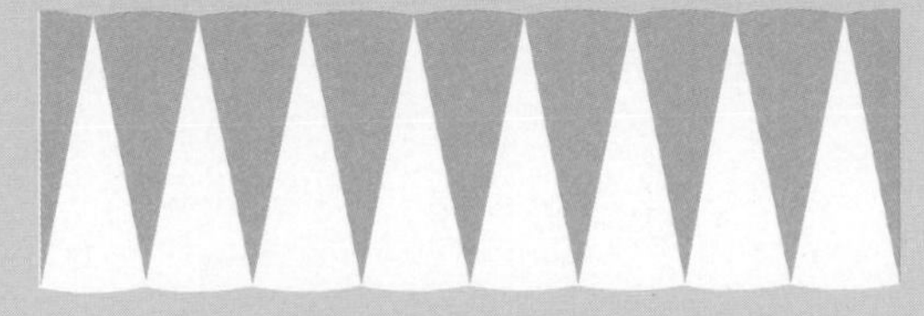

GOING EXPONENTIAL

The pattern of the basic number line has equal-sized increments—identical steps from 1 to 2 to 3, and so on. But in the seventeenth century it was realized that by taking growing steps it would be possible to produce a number line with a pattern that overcame its greatest practical difficulty—multiplication is tedious, while division is painful. (Anyone remember long division before the use of calculators?)

Multiplication on a conventional number line is a pattern of repeated addition. To multiply 5 by 4, we move up the number line by 5 lots of 4 (or 4 lots of 5). But not only is this slow, it becomes tricky when moving beyond whole numbers. It's easy to add, say, 3.7 to 4.2, but multiplying 3.7 by 4.2 using repeated addition requires a stretch of the imagination.

In 1864, English mathematician John Napier came up with a solution to this problem, which he called logarithms. A logarithm transforms a number into a different value, based on the pattern of an alternative number line. On the logarithmic line, the gaps between each number become smaller and smaller.

The starting point of a logarithm is its base, usually 10, 2, or a constant of nature known as e (chosen because it is useful in calculus). For simplicity, let's consider base 10. The logarithm (log for short) to base 10 of any number is the power to which you raise 10 to get that number.

WORKING LOGARITHMS

10^1 is 10, the logarithm of 10 is 1.

10^2 is 100, so log (100) is 2.

On a logarithmic number line, the first division to the right of 0 is 10, the second is 100. This is literally growing exponentially—it grows with the *exponent*—the power to which 10 is raised.

To get other values we move away from integer exponents. For example, $10^{1.5}$ is around 31.6. Every number bigger than 0 can be produced using some exponent of 10. The reason Napier came up with this strange number line is that to multiply two numbers, we simply add their logarithms. To divide them, we subtract their logarithms.

Remember that 10^1 is 1, while 10^2 is 100.

To get 100 we multiply 10 x 10.

But to get from 10^1 to 10^2 we add 1 + 1 = 2.

Similarly, 1,000 is 10^3, when we multiply, say 10 x 100, i.e. $10^1 \times 10^2$, so we add the exponents 1 and 2 to get 3.

A good way to imagine the pattern here is that the exponent gives us a number of dimensions. Going from 1 to 2 goes from a line to a flat shape. Going from 2 to 3 takes us to a three-dimensional form.

We can imagine that 0 dimensions is a point and this helps us understand what the logarithm of 1 has to be. Multiplying something by 1 leaves it the same. Adding the exponent 0 to another exponent leaves it the same. So, 1 is 10^0. Going to 4 dimensions and more, or to negative or fractional dimensions, is harder to envisage, but it is no problem for the mathematical pattern. Numbers smaller than 1 are achieved by dividing 1 by 10^n, represented as 10^{-n}.

Using this different pattern on the number line made calculations far easier and logarithms were used until replaced by calculators in the 1970s. "Logarithmic scales" are still common in scientific charts to represent dramatically changing numbers. Shortly after Napier published his book on logarithms, another English mathematician saw a way to migrate this new number line from the theoretical world to a pattern on a physical object.

Anyone old enough to remember the days before electronic calculators is likely to be familiar with the slide rule: a badge of honor among engineers of the period. This dates back to the 1620s, when William Oughtred put logarithmic number lines onto a pair of wooden rulers and slid one against the other to perform calculations.

"LOGARITHMS, THOSE NUMBERS SO IMPORTANT BY DIMINISHING THE LABOR OF TEDIOUS CALCULATIONS.

THOMAS THOMSON

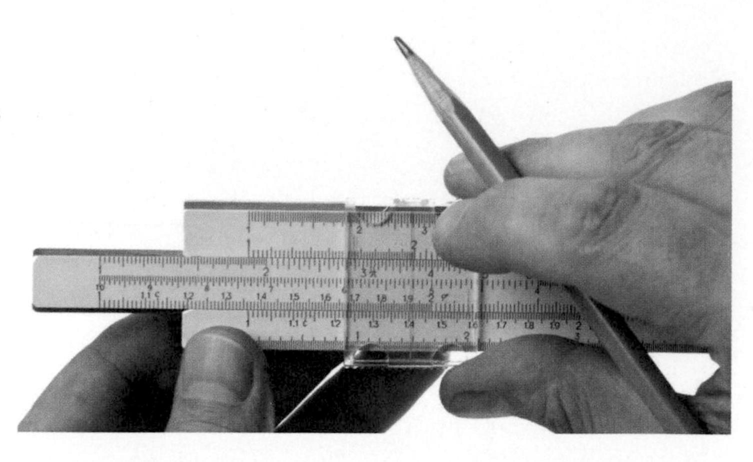

The slide rule brought the power of logarithms to quick, if approximate, calculations for engineers and scientists.

The slide rule had two fixed logarithmic rulers with a third placed in a slot between them, so it could slide back and forth. Over the top was a transparent slider, used to place a straight-line cursor across the sections.

It is the interaction between the number line patterns on the different sections that makes the slide rule work. To achieve the 3.7 x 4.2 calculation mentioned earlier, the center section would be slid until the number 1 on that section lined up with 3.7 on the top section. The cursor would then be moved to line up with 4.2 on the center section's scale and the answer 15.5 would be read on the bottom scale. The outcome was approximate, but good enough for practical use, and far quicker than working out the sum by hand.

"WHERE THERE IS LIFE THERE IS A PATTERN, AND WHERE THERE IS A PATTERN THERE IS MATHEMATICS.

JOHN D. BARROW

IMAGINARY NUMBERS

Even when used exponentially, the number line itself is a one-dimensional pattern, for example 1, 10, 1,000, ... However, when describing physical processes it is often useful to extend that pattern into two dimensions, for example when describing the behavior of a wave. This proved possible by bringing in imaginary numbers.

Mathematicians don't work in the physical world; in the mathematical universe anything is possible, as long as the mathematician follows the rules. We saw this happening with negative numbers, but they were just the start of excursions that take mathematics further from parallels with reality. Yet, perhaps surprisingly, some of those excursions prove extremely valuable.

Imaginary numbers are the result of asking the question: "Which number, multiplied by itself, produces a negative number?" In reality, there isn't one. If you multiply a positive number by itself you get a positive one; and if you multiply a negative number by itself you get a positive one. So, mathematicians dreamed up a number that multiplied by itself produces -1. This is represented by i. Any other imaginary number can then be produced, for example $2i$ or $-0.4i$.

Initially, imaginary numbers were a mathematical novelty. However, it was realized that they form their own number line, with a set of numbers that are totally detached from the real number line. This can be represented in two dimensions with the number lines

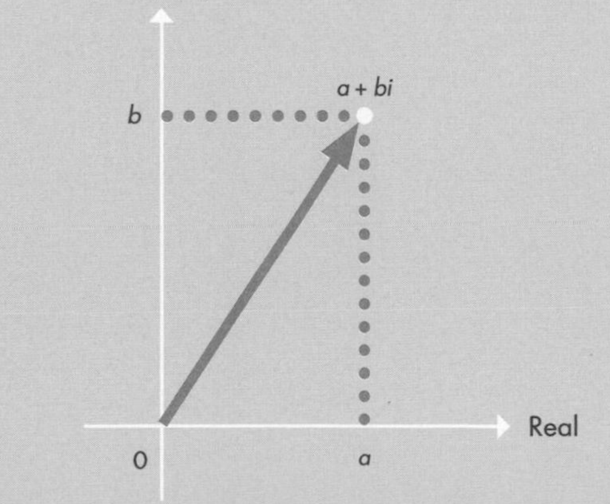

Complex numbers

By plotting a real number on one axis and an imaginary number on the other, a complex number represents a point on a surface.

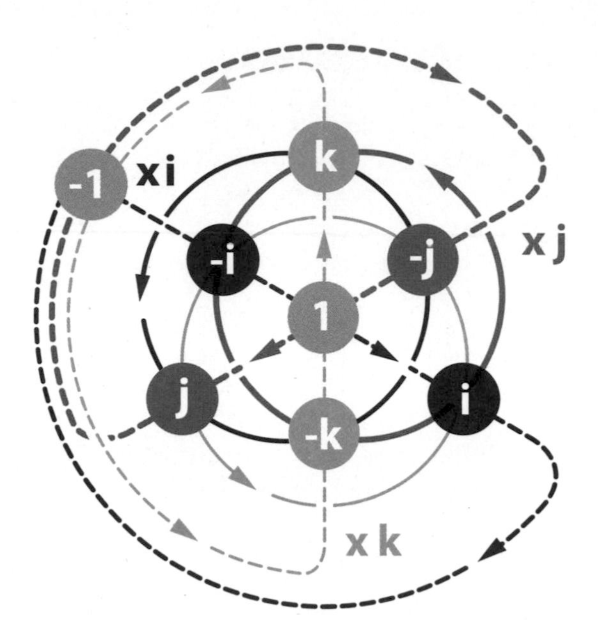

Quaternion multiplication

Although quaternions provided a valuable extension of complex numbers, manipulating them was difficult, as shown in this map of how the different parts interact in multiplication.

at right angles to each other, crossing at 0. Now we have a pattern covering every point on a two-dimensional surface, represented by a "complex number," combining a real and an imaginary number, for example, $3 + 4i$ or $2 - 3i$. Complex numbers are ideal for representing waves, so are widely used in physics and electrical engineering.

In the nineteenth century, the concept was extended to take in a potentially even more useful pattern called the quaternion, involving three sets of imaginary numbers plus a real one, producing a four-dimensional pattern. This feels like mathematicians having pointless fun; and they are happy to use thousands of dimensions to perform calculations if it helps. However, quaternions represented three spatial dimensions and one of time, reflecting many real-world physical processes.

Quaternions proved difficult to manipulate without computers (the diagram above shows the complex pattern of manipulations required to multiply quaternions) and the idea was pushed aside by a more flexible multidimensional approach called vector calculus, which is still used today. However, quaternions demonstrated the way that number lines can go even further beyond a single real dimension and still be valuable.

FOLLOWING FIBONACCI

One of the best-known variants on a number line is the Fibonacci series, named after the twelfth-century Italian mathematician Leonardo of Pisa, nicknamed Fibonacci, a contraction of *filius Bonacci* (son of Bonacci). Fibonacci's biggest contribution was the popularization of a change to the pattern used to indicate numbers in the West. At the time, the clumsy Roman system was the standard, where, for example, adding XIV to VI (14 + 6) was non-trivial. Fibonacci showed the flexibility and power of the Hindu system, already in use in Arabic countries, which included zero (non-existent in Roman numerals), symbols for 0 to 9, and the use of position to indicate powers of 10.

However, in the same book, *Liber Abaci* (Book of Calculation), Fibonacci produced an odd-looking variant on the pattern of the number line, now called the Fibonacci sequence. Rather than the familiar sequence 0, 1, 2, 3, 4, . . . this number line runs 0, 1, 1, 2, 3, 5, 8, 13, 21, . . . each number produced by adding the two previous numbers together.

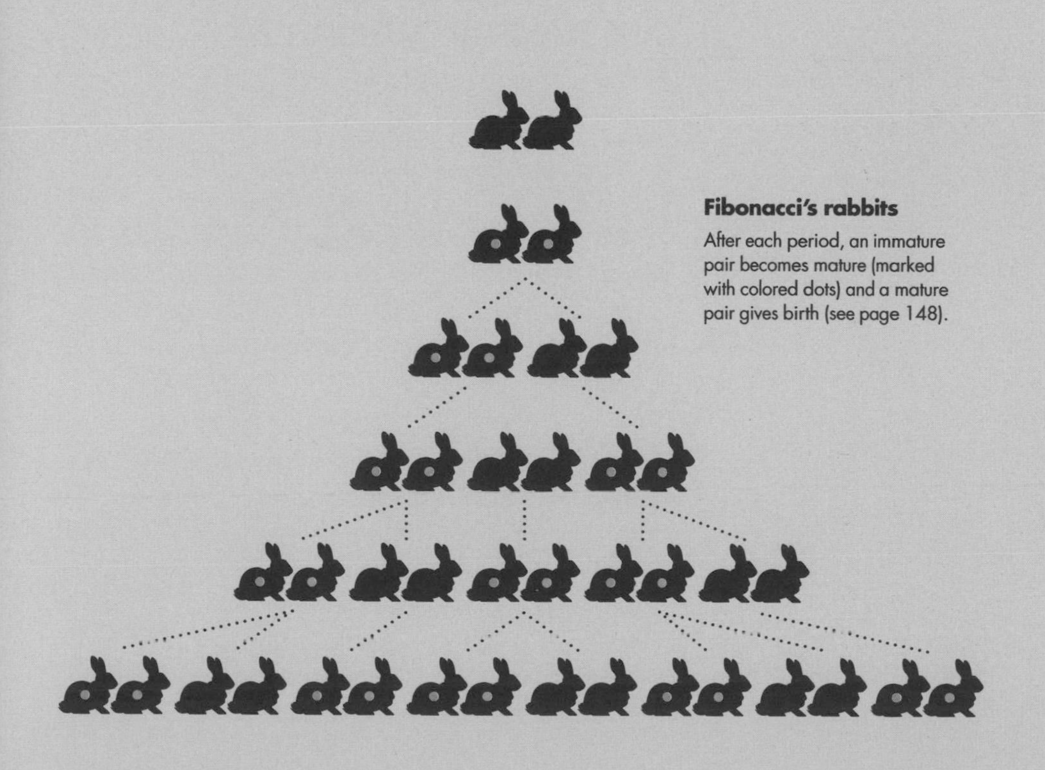

Fibonacci's rabbits

After each period, an immature pair becomes mature (marked with colored dots) and a mature pair gives birth (see page 148).

Fibonacci introduced the sequence as the pattern produced by breeding rabbits (see page 147). These rabbits take one month to mature. Each pair produces a mixed sex pair every month, once they are mature, and no rabbits die. Although not realistic, this was a first attempt to apply a mathematical pattern to a population. In the first month we have a pair of immature rabbits: one pair. Next month they become mature: still one pair. The following month they produce babies, so now there are two pairs. A month later another pair is produced, and the first babies mature. We now have three pairs. Next month we get five pairs, and so on.

Although the series can't model real populations, it does turn up in nature. This pattern reflects well the way some biological processes develop through time, for example, the distribution of seeds in the head of a sunflower follows a Fibonacci sequence. The same visual pattern is made more geometric by plotting the sequence into two dimensions producing an approximate "golden spiral," which grows in the same proportion each quarter turn. This pattern occurs in nature, shown in both the shell of the chambered nautilus and the proportions of spiral galaxies.

A KNOTTY PROBLEM

DONUTS AND CUPS:
In topology, objects are identical if one can be stretched and distorted into the shape of the other without tearing. As a result, a donut is topologically identical to a cup with a handle.

From geometry we can move the number line patterns into the mathematics of the distortion of multidimensional shapes, known as topology. One such study is knots, producing a new, distinctive number line variant. Mathematical knots are unlike real world knots in that the imaginary "string" has no ends. Knot theory produces a rapidly growing number line showing the numerous different patterns that can be produced for any required number of "crossings," (this is the number of times the string in the knot passes over the other strings).

The knotty number line runs 1, 0, 0, 1, 2, 3, 7, 21, 49, 165, . . . crossings. Here, the pattern emerges from the topological requirements of operating in three dimensions. The sequence begins with an oddity (the "unknot") a loop of string with no crossings. It is then impossible to find a way for the loop to cross itself just once or twice; the simplest true knot is the so-called trefoil, with three crossings. There is only one way to arrange this, as is the case for four crossings. After that the number of arrangements increases rapidly.

Knots and crossings

Some of the simplest mathematical knots.

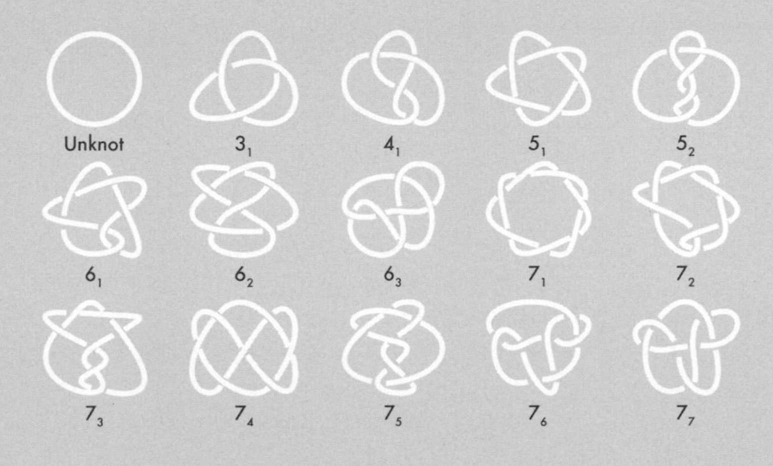

Initially knot theory had no practical value—it was of mathematical interest, but with no application. The nineteenth-century Scottish physicist Lord Kelvin did try to apply it to explain atoms as knots in the aether—an invisible substance thought then to fill all space—but this proved fruitless. It remains primarily a theoretical concept, although it does have some use in molecular biology.

DNA, the molecule containing the pattern of life (see Chapter 9), is a very long molecule, which in many cells spends most of its time tightly wound up. When a cell divides, the DNA has to be unraveled, split in two, and each half built back up to form a pair of identical molecules. This process involves chemicals called enzymes cutting through, unknotting, and reconnecting parts of the molecule. Knot theory has helped biologists gain insights into this process.

“ IN A KNOT OF EIGHT CROSSINGS, WHICH IS ABOUT THE AVERAGE-SIZE KNOT, THERE ARE 256 DIFFERENT ‘OVER-AND-UNDER’ ARRANGEMENTS. ANNIE PROULX

GET SET GO!

The simplicity of number lines is also their power, reflecting their basis in the pattern that lies beneath both numbers and arithmetic: set theory.

In mathematics, a set is a particular type of pattern: a group of things that share one or more properties. In the physical world, you might have a set of objects that are colored orange, or a set of things that you noticed this morning. In mathematics, sets usually have numeric properties. So, for example, there is the set of the integers, or the set of the prime numbers. We can often partition part of a set. This forms a set in its own right, a subset of the original set—a term that has moved into general usage. So, for example, the set of the even integers is a subset of the set of integers.

Sets are used to define the integers in a way that builds in a repeating pattern like Russian dolls, each definition slotted one within the other. Zero is represented by the empty set: a set with nothing in it at all, indicated mathematically as Ø. One is the set containing just the empty set: { Ø }. Two is the set containing the empty set and the set containing the empty set { Ø, {Ø} }, . . . and so on.

Set theory is the foundation of arithmetic and many other aspects of mathematics, underpinning the number line. We don't need to know the details of set theory, but one useful aspect is the cardinality of the set, which describes its size. This is valuable because we use cardinality to work out if two sets are the same size, even if we don't know how many items there are in the sets. Two sets have the same cardinality if we can pair off all the members of the first set, one by one, with the members of the second set.

A simple example would be the set of the seasons and the set of legs on a dog. You can pair off one season with each leg, so you know that the two sets have the same cardinality without knowing the number of legs. Cardinality is also how we use numbers to stand in for real things. If you have eight oranges, for example, what you mean is that you can pair off one orange with each of the set of integers from 1 to 8, so the set of oranges has the same cardinality as this subset of the integers, which we condense to: "You have eight oranges."

Cardinality is essential when dealing with infinite sets. Remember, infinity isn't a number, and we usually deal with potential infinity, but if we consider every positive integer, say, on the number line we are dealing with a truly infinite set. The cardinality of this set is given a different symbol to the lemniscate—it is $\aleph_0$—pronounced aleph null. (Aleph is the first letter in the Hebrew alphabet.)

Knowing about sets and cardinality means that we can understand better how number lines work when considering the entire line, rather than a part of it. Some subsets of the positive integers are finite, for example, the set of integers from 3 to 47 (or any other pair of numbers). But other subsets are infinite, for example, the even positive numbers.

A more dramatic demonstration is the set of squares of the positive integers. The integers go 1, 2, 3, 4, . . . while the squares go 1, 4, 9, 16, ... With our finite-minded approach there appear to be far more integers than squares, for example, all the numbers like 2, 3, 5, 6, 7, 8, 10, 11, 12, 13, 14, 15, ... Yet, because we are dealing with sets, we need to consider cardinality. The set of positive integers can be put in a pattern of one-to-one correspondence with the squares. There's a square for each integer. As a result, they have the same cardinality. One of the properties of an infinite set is that it will have subsets with the same cardinality.

1 → 1
2 → 4
3 → 9
4 → 16
5 → 25
6 → 36
7 → 49
... → ...^{2}

Cardinality squared

Each positive integer can be paired off with one of the squares, meaning that the set of the integers and the set of the squares have the same cardinality.

UMBRELLAS ON THE NUMBER LINE

Once we have the concept of cardinality and infinite sets, we can reveal one of the fascinating paradoxes of the infinite number line: the attempt to keep the number line dry. The starting point is the set of every rational fraction; that's every number you can write as one integer divided by another integer. A German mathematician called Georg Cantor proved that this set has the same cardinality as the integers. Cantor envisaged a table of every possible fraction, then provided a step-by-step pattern to work through it. As long as you can set up such a pattern, you can put each item in the table in one-to-one correspondence with the integers as they have the same cardinality.

Let's look at a set we have already met: the set of fractions that are 1 divided by the powers of 2: ½, ¼, ⅛, 1/16, ... Again, we can pair these off one-to-one with the integers, so they have the same cardinality. But remember that we found that the sum of the infinite series 1 + ½ + ¼ + ⅛ + 1/16, . . . was 2, so the total of all the fractions mentioned above is just 1.

Table of fractions

For each subsequent column the top number increases by one and for each row the bottom number increases by one (this table holds every possible fraction). Some are repeated, the diagonal values are all 1.

1/1	2/1	3/1	4/1	5/1	6/1	7/1	8/1	9/1	10/1	...
1/2	2/2	3/2	4/2	5/2	6/2	7/2	8/2	9/2	10/2	...
1/3	2/3	3/3	4/3	5/3	6/3	7/3	8/3	9/3	10/3	...
1/4	2/4	3/4	4/4	5/4	6/4	7/4	8/4	9/4	10/4	...
1/5	2/5	3/5	4/5	5/5	6/5	7/5	8/5	9/5	10/5	...
1/6	2/6	3/6	4/6	5/6	6/6	7/6	8/6	9/6	10/6	...
1/7	2/7	3/7	4/7	5/7	6/7	7/7	8/7	9/7	10/7	...
1/8	2/8	3/8	4/8	5/8	6/8	7/8	8/8	9/8	10/8	...
1/9	2/9	3/9	4/9	5/9	6/9	7/9	8/9	9/9	10/9	...
1/10	2/10	3/10	4/10	5/10	6/10	7/10	8/10	9/10	10/10	...
...	...	...	...	...	...	...	...	...	...	...

Cantor's route

A step-by-step route through the table of the fractions. Because this can be set up, the fractions and the integers have the same cardinality.

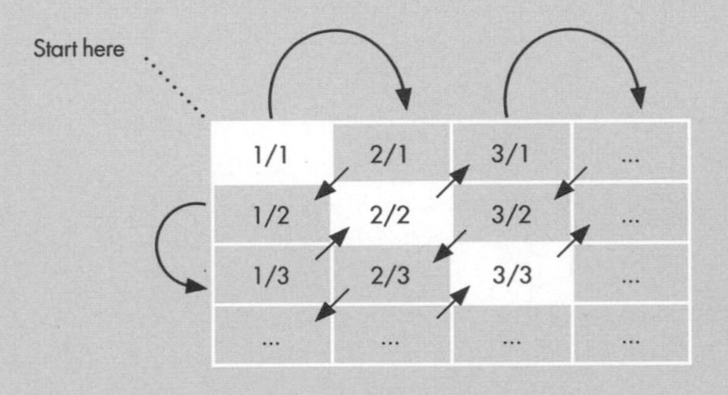

Imagine we wanted to protect the whole number line from getting wet. We are going to issue each rational fraction on the line with an umbrella. The umbrella will be a simple T shape. The first umbrella we give out is ½ a unit of the number line across the T. The second umbrella given out is ¼ of a unit of the number line across, and so on. Once every rational fraction has an umbrella, the whole number line is covered. The umbrella extends half its width in either direction, so, for instance, the first umbrella will cover all numbers for ¼ of a unit to its left and ¼ of a unit to its right. In each direction, the umbrella reaches another rational fraction.

Now bearing in mind we've issued an umbrella to *every* rational fraction; the whole number line is covered by this pattern of umbrellas. But, remember how wide the umbrellas were. Their widths form the infinite series ½ + ¼ + 1/8 + 1/16. . . . So, the maximum amount of the number line those umbrellas can cover is 1 unit. A set of items with a total width of just 1 manages to cover a line that goes all the way to infinity. This is why infinity is mindboggling.

The number line starts off as a simple tool for addition but represents far more. The way something apparently simple can transform our understanding also applies to our next pattern called a cladogram, which reflects the genetic structures of the natural world.

8
CLADOGRAMS

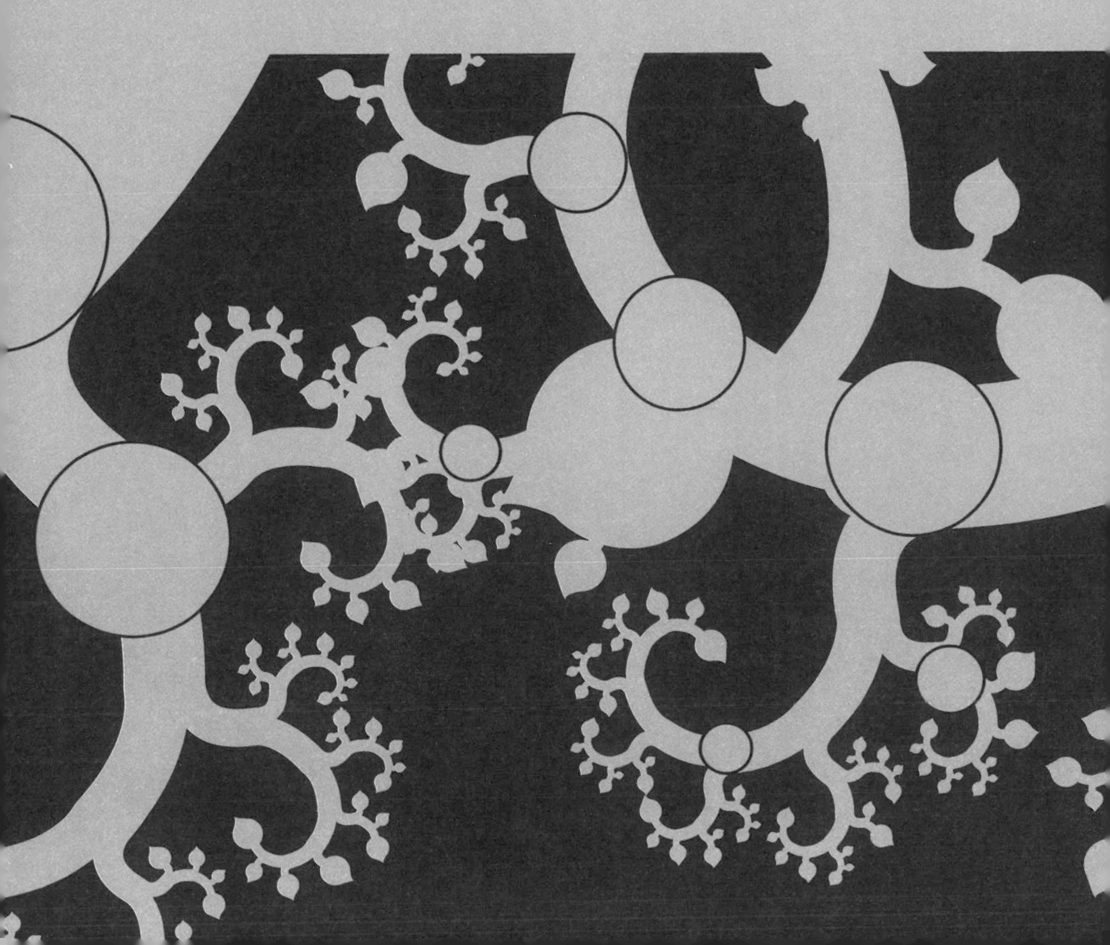

Charles **Darwin**
1809–1882

THE TREE PATTERNS OF EVOLUTION

When Charles Darwin's great work on evolution, *On the Origin of Species*, was published in 1859 it had a single illustration. The treelike diagram, spreading upward from the bottom of the page, showed how similarities between species could imply common ancestors. That original was a sketchy series of lines, but as the fundamental nature of evolution spread, so did the sophistication of such diagrams. Some, such as those drawn by the German naturalist Ernst Haeckel took the form of literal trees with labeled branches—others were far more abstract. But all were designed to make clear relationships between species. Such trees are still useful, but our understanding has been transformed by the cladogram, a later form of tree diagram, which shows how species have formed through time, based on genetic information. Cladograms clarify the pattern of evolution.

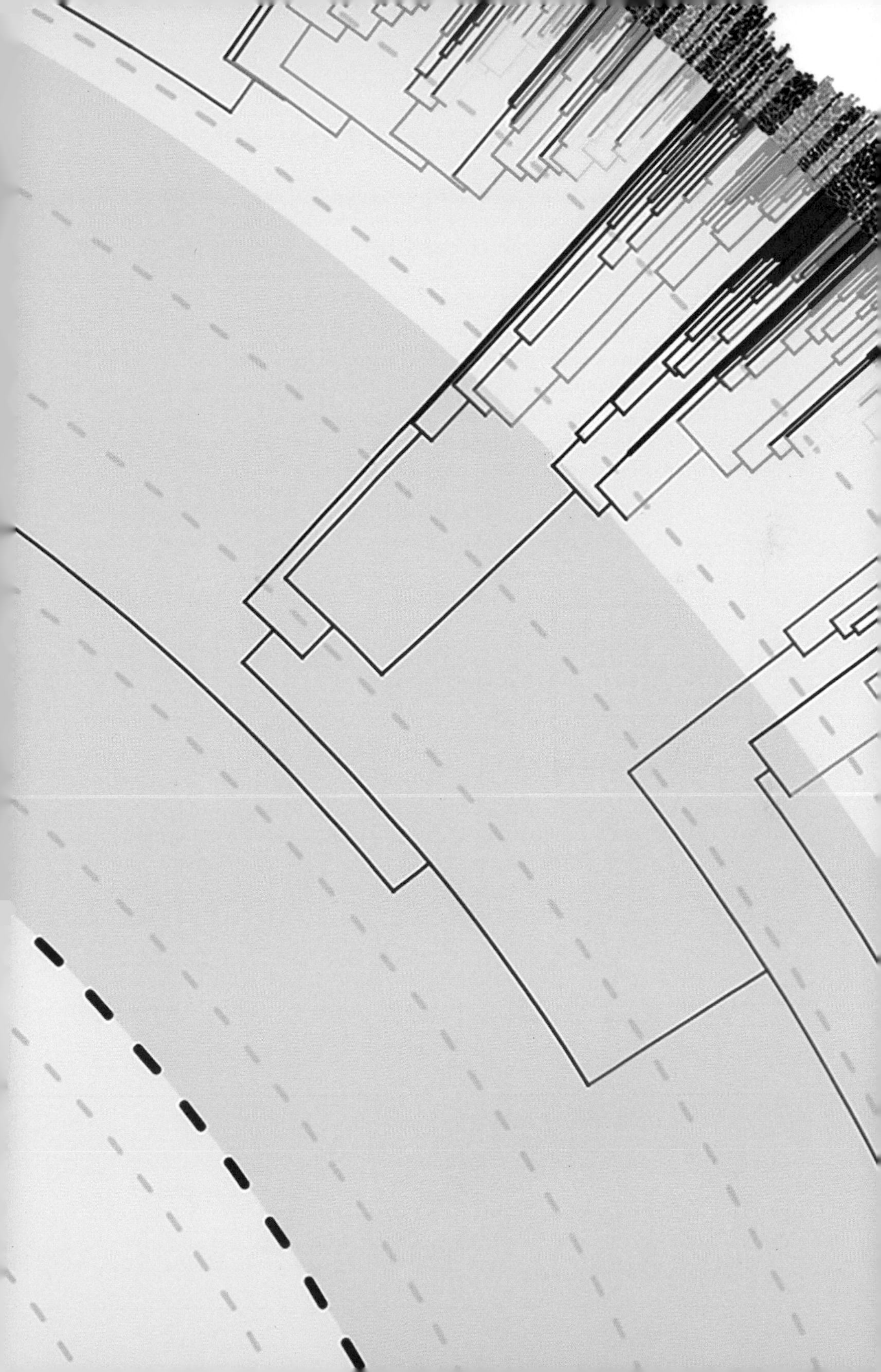

THE EVOLUTION OF DARWIN

To understand the essential diagram that represents the pattern of natural history, we need first to discover how the concept of evolution that lies beneath it originated. The cladogram is inherently driven by the way that different species have evolved from common ancestors. For Charles Darwin, it all began with beetles and rocks. Darwin's parents had intended him to be a doctor, but after two miserable years at medical school in Edinburgh, Scotland, Darwin switched to Cambridge University in England. Here, he had intended to specialize in theology, but a fascination with insects and the new understanding of the age of planet Earth pushed him in an entirely different direction.

At the start of the nineteenth century, the prevailing assumption had been that Earth had been created around 6,000 years earlier, and that changes in Earth's surface had been down to catastrophic events, notably the Biblical flood. However, a newer concept—uniformitarianism—portrayed a world that had gone through many gradual changes over millions of years.

UNIFORMITARIANISM: The idea that Earth's structures were the result of long-term, continuing processes, rather than the single, drastic event of the opposing catastrophism.

Darwin's tutor, the biologist and geologist John Henslow, had been asked to travel as a naturalist on the ship HMS *Beagle*, which in 1831 was to leave for South America on a five-year voyage of scientific and geographic exploration. Henslow's wife was not keen on her husband going: Henslow persuaded the captain of the *Beagle*, Robert Fitzroy (who was responsible for the first weather forecasts), to accept the young Darwin in his stead. Darwin's experiences would provide the trigger for his development of the theory of evolution.

The anole lizard has demonstrated rapid evolutionary change in cities, where it has developed stickier feet, which are better able to climb walls and windows.

A male zebra and a female pony can produce offspring known as zonies. Unable to reproduce, such hybrids are not true species.

At the time, the common understanding was that all existing animals and plants had been in place unchanged since the Christian concept of "creation." The fossilized remains of unfamiliar animals that were found were thought to be from organisms that had died out in the flood, hence the frequent reference to dinosaurs as being antediluvian—literally "before the flood." This was a short-sighted view. Already, there was plenty of evidence that selective breeding could bring about dramatic changes in domestic animals and plants. It perhaps shouldn't have been such a great leap that natural influences could have a similar effect, producing the same impact through natural selection as selective breeding did under human intervention.

FOSSILIZATION: The formation of fossils can occur in several ways. Mostly commonly it happens when water containing minerals seeps into the dead organism, replacing soft parts as minerals harden.

Without the concept of different animals and plants evolving away from their ancestors, the idea of a tree of species would be meaningless. The pattern would be nothing more than a set of parallel lines. However, on his voyage Darwin was able to study the way that animals and plants differed around the world. In particular, he would see how isolated islands could make it impossible for a species to migrate to a better suited environment, meaning that if a species was unable to adapt, it would not long survive.

EVOLUTION REVOLUTION

A key inspiration for Darwin would be the way that the variation of species of birds and tortoises from one to another of the Galápagos Islands suggested that, once isolated on an island, a species would gradually change to have the best chance of survival in that particular environment. This was in opposition to the thinking of the time, which assumed species were fixed in creation and never changed.

ALFRED RUSSEL WALLACE: In 1858, Darwin received a letter from Welsh naturalist Wallace proposing a theory very similar to the one Darwin had been developing for 20 years. Rather than dispute who came first, they jointly published their ideas.

Although it took Darwin many years to publish his landmark book, within a year of his voyage he was already playing with a treelike pattern to map out how different species were related. Central to Darwin's revolutionary idea was the concept that organisms, which varied from individual to individual, could pass on characteristics to their offspring. If those characteristics benefited survival in a particular environment, the offspring were more likely to survive to reproduce and pass them on again, gradually transforming a species into a new one.

Darwin had half the picture. He knew that evolution through natural selection made a lot of sense. In order to work, though, it required three things to be true. Firstly, living in an environment where having particular characteristics made survival more likely. Nature provided this situation. Secondly, his theory required variation

Darwin's finches

Examples of different beak forms in Darwin's finch species, from Darwin's 1845 journal. The birds are:
1 Large ground finch
2 Medium ground finch
3 Small tree finch
4 Green warbler finch.

Darwin noticed a variation in the size and shell form of tortoises between the Galápogos Islands, depending on the environment. These tortoises, like this giant Galápagos tortoise, can live for over 100 years.

between different organisms in the same species. Finally, there was a need for a means of transferring characteristics from generation to generation.

A central piece of evidence for this idea was a group of around 14 finch species on the Galápagos, now known as Darwin's finches. Imagine that a change in the environment made most of the foodstuffs available for finches only accessible to those able to crack hard shells with big, strong beaks. If finches had differing beak sizes, then the finches with bigger beaks were more likely to survive and have offspring. If the offspring of those finches were also more likely to have beaks bigger than in the general finch population, then over time more and more of the finches would be of the big-beaked variety.

CHARACTERISTICS: Features of a living organism that may influence its ability to survive. Can be anything from beak size to brainpower.

The dramatically colored sea iguana is found only on the Galápagos. It evolved for this niche environment where it feeds on sea algae.

We now know that the mechanism behind the second and third requirements is genetics: the variation between organisms and the transfer of characteristics from generation to generation. The detailed workings of this will come up in the next chapter when we examine that fundamental biological pattern, DNA. Here, it is sufficient to know that evolution can produce new species when genetic variation—encouraged by factors such as a changing environment—makes it possible to better survive.

“THIS PRESERVATION OF FAVORABLE VARIATIONS AND THE REJECTION OF INJURIOUS VARIATIONS, I CALL NATURAL SELECTION.

CHARLES DARWIN

THE RAINBOW OF EVOLUTION

Evolution is sometimes presented as being controversial, yet the basic concept of species changing through the generations to reflect environmental conditions is nothing more than common sense. It's not the science that is surprising, but rather that it took so long for evolution to become an established scientific theory.

However, there is one somewhat artificial pattern that still causes confusion, and that is the division of organisms into species. Even many who deny "Darwinism" accept that organisms are able to pass on characteristics that enable their descendants to survive better, but they argue that this just makes, for example, a lion into a better lion, or a bee into a better bee. They accept such evolutionary changes within species, but not that new species can evolve, a process that we will see mapped out in the patterns of cladograms (see pages 169 and 171).

The reason that those who doubt evolution have a problem is that the concept of "species" isn't really a scientific one. It's a concept from natural history, a field that was originally dedicated to cataloging organisms, rather than explaining their existence scientifically. In reality, although each organism is the same species as its parents, it is entirely possible to have a transformation of species across the generations. A useful analogy to understand this paradox is the familiar pattern of the rainbow.

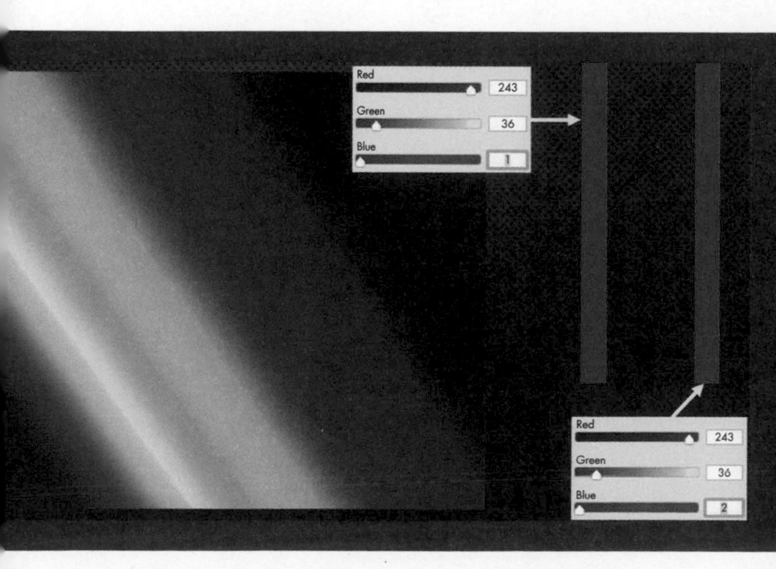

The rainbow analogy

A computer represents over 16 million rainbow colors using settings for red, green, and blue. The two vertical lines appear identical, despite having different RGB values, but across the range all the rainbow colors are shown.

We are usually taught at school that rainbows feature seven distinct colors (red, orange, yellow, green, blue, indigo, and violet). This is simply not true. It's a fiction that was dreamed up by Isaac Newton, probably because he wanted there to be seven rainbow colors to parallel the seven musical notes. In reality, the rainbow contains billions upon billions of colors. Even on a home computer, the full color spectrum available has a total of 16.7 million colors to choose from.

Imagine each of those many colors is a single generation in a tree of organisms. If we pick out any two adjacent colors from the 16.7 million, they will be indistinguishable. As far as our eyes are concerned, they are exactly the same color. Similarly, two generations of an organism are the same species. Any two. But if we look far enough through the color spectrum, we will go from red to orange to yellow, and so on. In the same way, as we pass through many generations, we will reach totally different species.

Each individual is the same species as both its parents and its offspring, yet each generation has small genetic differences. Over time, these accumulate until the difference is large enough to have a new species. Over the 4 billion years or so that there has been life on Earth this process has enabled the changes that have occurred from a single original life-form to every organism still existing on the planet.

The dog dramatically demonstrates the result of conscious selection from generation to generation on desired characteristics. Eventually a different species may result, but as yet dogs remain the same species.

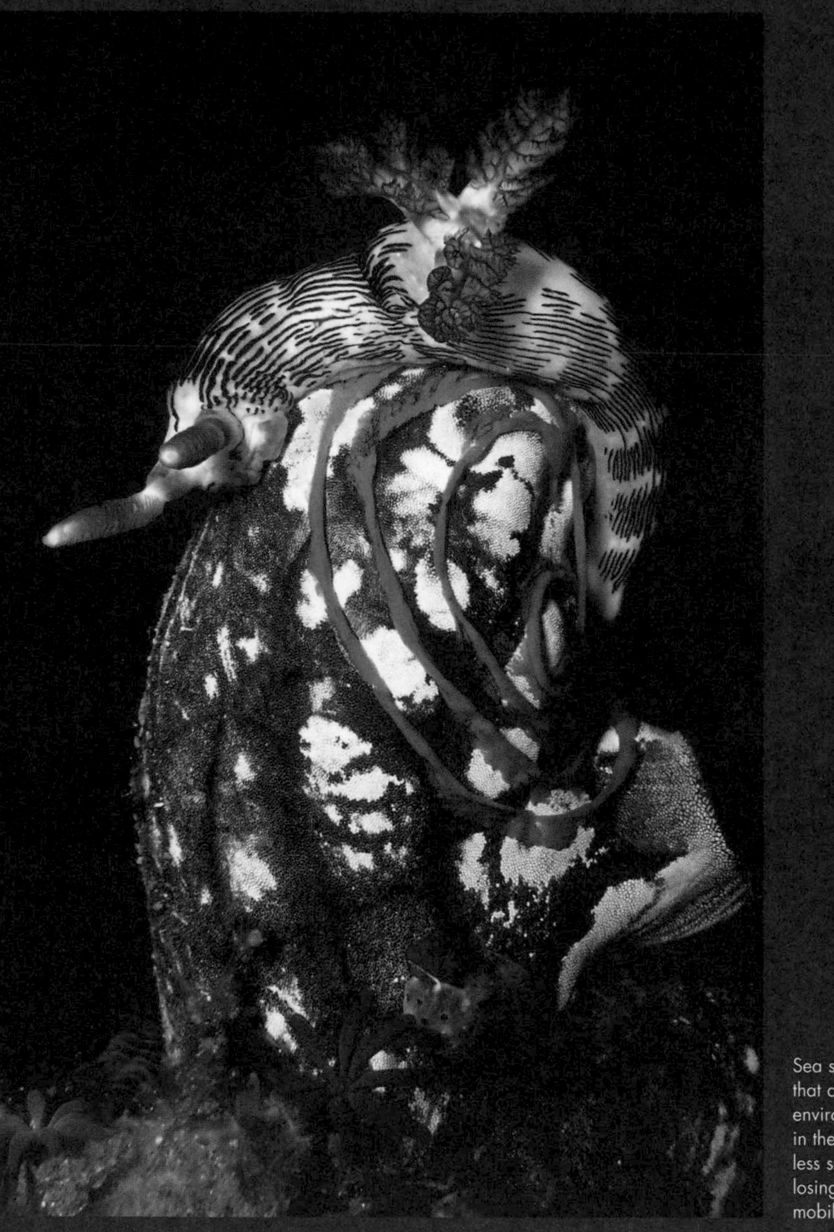

Sea squirts demonstrate that a response to the environment can result in the development of a less sophisticated form, losing the organism's mobility and brain cells.

A PATTERN OF RANDOMNESS

There is a classic drawing, which has been reused in many different ways, that shows the imaged pattern of the ascent of humans. It portrays a chain of development as evolution drives humanity toward the goal of a fully evolved *Homo sapiens*. As such, it is total nonsense, because evolution is not driven by a goal. There is no preset pattern to evolution, but rather a distinct absence of pattern: randomness.

Those who are uncomfortable with evolution are often also uncomfortable with the idea that something can happen purely by chance. It comes back to the concept we've already mentioned: that human beings are pattern-seeking in order to be able to deal with the world around us. We don't just accept that things happen, we want there to be a reason for them to have happened. When things go wrong, we look for someone or something to blame. When things go right, we praise the individuals involved, rather than considering them to be the result of luck.

Of course, this doesn't mean that everything lacks a cause and is purely random. In one sense everything does indeed have a cause, but that cause is often not driven by a rational mind that has a goal, making something happen by intent. Far more of what happens brings together a whole range of random, or near-random inputs, producing the kind of messy, complex environment that distinguishes the real world from our idealized plans and models of it.

With our pattern-driven, outcome-focused minds, it can be easy to see evolution as something with a particular long-term direction or goal in mind, but nothing is further from the truth. There *is* a direction in terms of surviving current environmental conditions, but there is no big picture, and nothing about evolution has some big picture of what it means to get better in an abstract sense. It's all about short-term survival. You could call it a form of directed randomness; there is no goal, simply a series of local preferences.

We can see the way that we need to move away from our human-centered ideas of what is an improvement in the example of the sea squirt. These simple aquatic organisms start life as swimming larvae, not unlike tadpoles, which have a basic brain and light-sensing organs to guide them. As they mature, sea squirts go through a metamorphosis, attaching themselves to a rock where they will remain for the rest of their life. Not only do they lose their equivalent of eyes in this process, even their brain cells are absorbed and recycled, leaving the organism only with the basic nervous system that is required for it to function. The sea squirt makes the changes needed to survive, but in doing so it becomes less sophisticated. Similarly, the pattern of evolution is not one of seeking ever greater sophistication and refinement; if survival is aided by becoming simpler, then evolution will happily head that way. Evolution takes a random walk, but the options available to it are shaped by the environment.

HOMO FLORESIENSIS

Closer to home, perhaps the best example of why that diagram showing a directed human chain of evolution misses the point is a form of life that flourished on the island of Flores in eastern Indonesia, about 60,000 years ago. As recently as 2004, the fossilized remains of a new humanoid species was discovered on Flores. These extinct hominins would be nicknamed "hobbits" as they were only about 3 feet (1 meter) in height. Their proportions were not unlike humans, but they had particularly small brains, even compared with hominins from earlier species.

HOMININS: The name given to humans, chimpanzees, and a range of extinct earlier life-forms that share some human-like characteristics.

Note that date of around 60,000 years ago. We know that our own species, *Homo sapiens*, has been in existence for about 200,000 years—several times longer. Therefore, the "hobbits" were still in existence when humans were already common on Earth. They evolved before *Homo sapiens* did, quite possibly from larger ancestors than their compact form. What often seems to happen on isolated islands is that the evolutionary response to the limited habitat is to favor relatively small animals. There have been examples, such as pigmy hippopotami and elephants, fossils of which have been found in island environments. It seems likely that the diminutive *Homo floresiensis* evolved from a hominin species that was larger in size, with bigger brains. The species evolved to be smaller and to have less of a brain because that best suited survival on Flores.

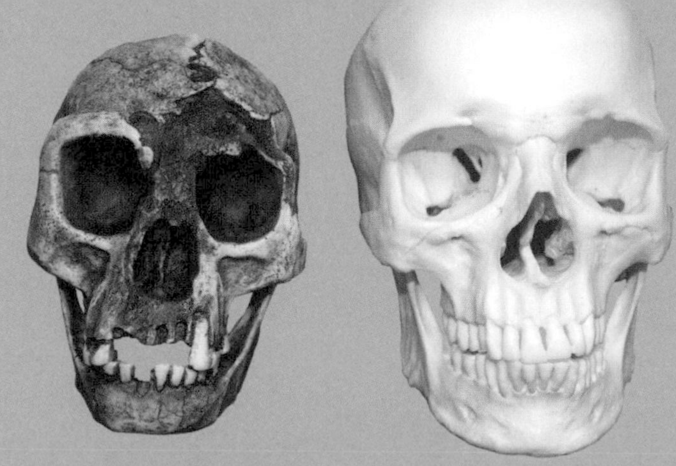

A comparison of a human skull (right) with one of *Homo floresiensis* (left) shows the significantly smaller size, particularly in brain capacity.

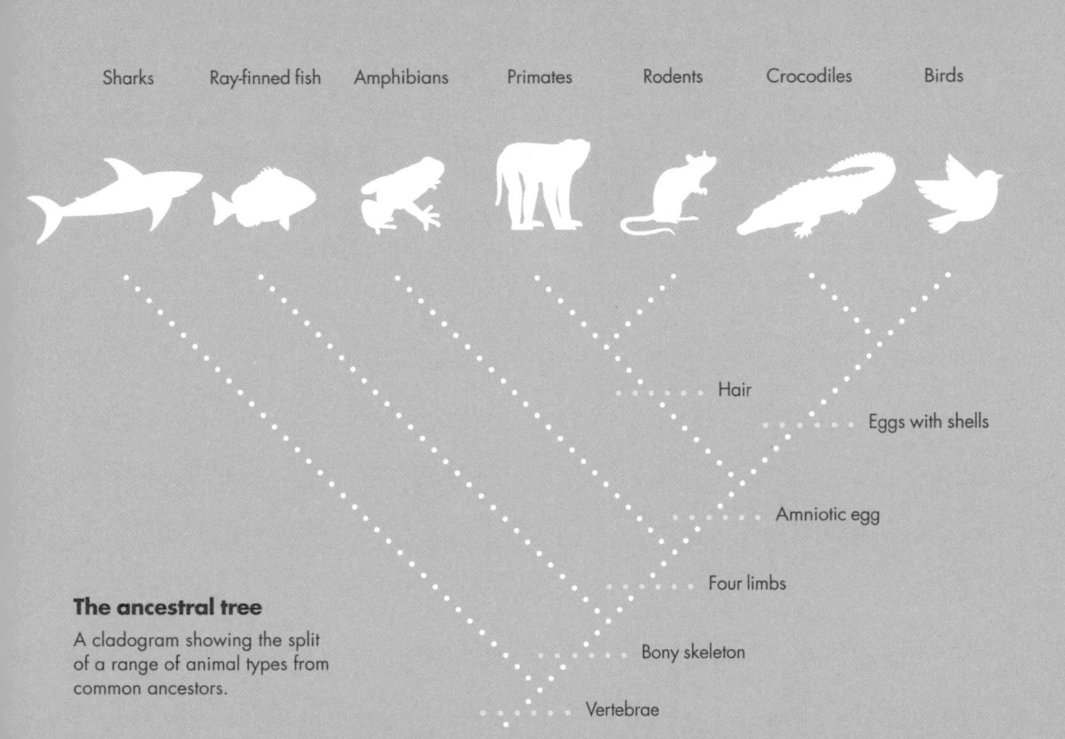

The ancestral tree

A cladogram showing the split of a range of animal types from common ancestors.

THE ESSENTIAL COMMON ANCESTOR

Cladograms are patterns that make clear our best guesses of how we can trace back linkages between species. Although structurally different, they are a bit like a family tree in concept, in that they work by linking together different species based on a common ancestor. This concept of a "common ancestor" is an important correction to our tendency to see ourselves as, in some sense, the destination of the pattern of evolution. We need to be reminded that we aren't the only organisms that have evolved.

Historically, it was not unusual, for example, to think of the other great apes as our ancestors. Infamously, in a public event on the topic of evolution at Oxford University's Museum of Natural History this fallacy is said to have cropped up in an exchange between Bishop Samuel Wilberforce and the biologist and Darwin-supporter Thomas Huxley. There is sadly no precise record of what was said in the debate, but Wilberforce is reported to have asked Huxley something along the lines of whether he would rather he was descended from a monkey (or gorilla) on his grandmother or grandfather's side.

In reality we are not an evolutionary development from any of the other great apes (chimpanzees, bonobos, gorillas, and orangutans). The cladogram pattern shows us that each species split off from a common ancestor at a different point. First orangutans, then gorillas split off, with the divide between us and our closest relatives, chimps and bonobos, happening most recently. In each case, the change was from a common ancestor, not from the "more primitive" of the available apes. Admittedly, such a common ancestor is likely to have looked more like other apes than like us, because we are a particularly freakish ape in the way that we have little visible fur, but that doesn't make those shared ancestors any more a chimp or a gorilla than we are.

In fact, since our split from chimps and bonobos, they have evolved more than we have in terms of shared genetic markers. A study showed that of 14,000 matching genes compared between humans and chimpanzees, 233 chimp genes had changed through positive natural selection, whereas only 154 human genes had altered. Of course, we have gone way beyond the capabilities of evolution in a way that no other animal ever has. Where once our environment controlled our evolution, now we have radically transformed the environment (not all for the better).

Bonobos are close relatives of the chimpanzee, but with a far less aggressive social structure. They share a common ancestor with chimps, which shares a common ancestor with us.

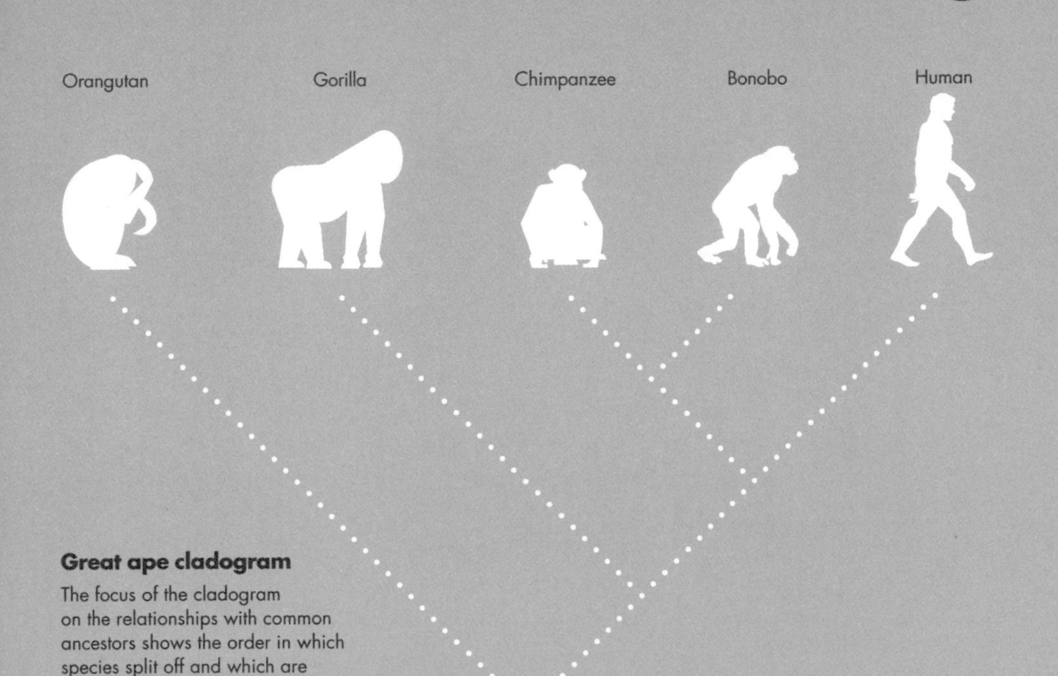

Great ape cladogram

The focus of the cladogram on the relationships with common ancestors shows the order in which species split off and which are closest relatives.

HOMININ FOSSILS

There is an important proviso in the previous section, saying that we evolved from a hominin species that was not our own. In the majority of cases, because of a lack of DNA, it is impossible to say whether hominin fossils are our direct ancestors or which common ancestor they evolved from. *Homo floresiensis* was originally thought to be a relatively recently evolved species, but we now believe it is significantly older than humans, splitting off long before many other hominins. Misunderstanding this, on a regular basis the media will carry some story claiming a new discovery is "mankind's oldest ancestor." This claim was made, for example, in 2019 for a species referred to as MRD.

In reality, all that had been found was a 3.8 million-year-old cranium from a hominin. But we do not know if this was an ancestor, or a species that occupies a branch from a common ancestor that does not lead to human beings. Without doubt, we will have a common ancestor—every living thing on Earth so far discovered appears to have had a common ancestor. But that doesn't put every old species in a direct line in our cladogram pattern.

We can, with reasonable certainty, draw a cladogram that puts us in relation to the other great apes and can roughly deduce how long ago each of those splits occurred. This is because we have plenty of DNA from the current great apes, and by studying the ways that genes differ between us it is possible to get approximate timings. But the vast bulk of the fossil record is missing.

Fossilization is a rare phenomenon, and fossil discoveries are chance occurrences. So, rather than having a neat cladogram pattern showing all our hominin ancestry, it is as if someone has redacted more than 90 percent of the tree, leaving us with a few scattered twigs and no way of seeing how they connect together. We can date a fossil reasonably accurately using radioactive dating techniques, but this tells us nothing about the relationship between species. We can provide a latest date for when a species split off from a common ancestor, but often cannot put it in a direct line of descent.

HALF-LIFE: The concept of "half-life" is most commonly applied to radioactive materials, where it describes the time for half the initial amount of the element to decay.

The further back we go in time, the harder producing a picture of relationships becomes. As we have seen, the best patterns of cladistics are driven by genetic information, dependent on extracting DNA from living organisms and old specimens and making comparisons. In fiction, we are used to the idea of DNA from the past providing dramatic opportunities from bringing extinct animals alive. *Jurassic Park*, for example, suggested dinosaurs could be recreated from DNA found in blood-sucking insects trapped in amber.

Unfortunately (or fortunately bearing in mind how the *Jurassic Park* movies inevitably turn out), this is never going to happen. DNA deteriorates over time. It has a half-life of around 520 years. This means that for every 520 years we go back into the past, around half of a DNA sample will have become useless. Go back around 1.5 million years or more and nothing useful will remain. The dinosaurs went extinct over 60 million years ago. So, dinosaur cladistics has to be based on indirect indications: a combination of the dating of fossils and the patterns of changes in skeletal structures, bone shapes, and more (so-called morphology) that indicate some underlying genetic transformation. The same is true with the older hominin specimens.

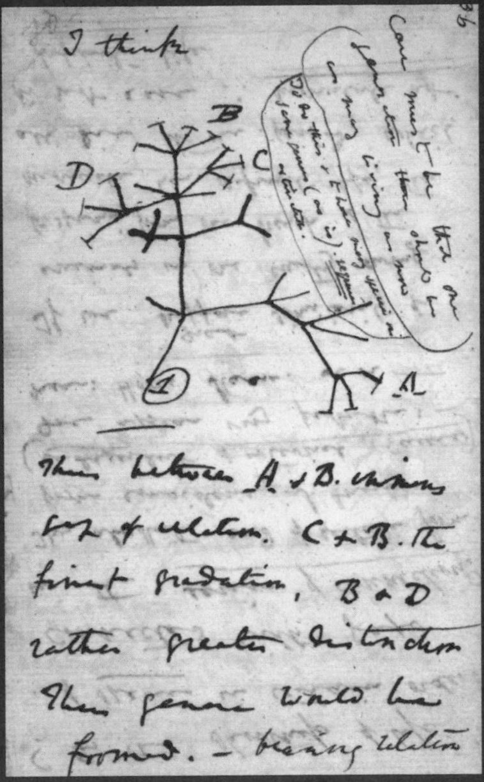

Darwin's first tree of life diagram, drawn in his circa 1837 notebook.

FROM LITTLE SHOOTS TO GREAT OAKS

Darwin first sketched a simple tree structure—in reality more like a giant hogweed plant than a tree—in his notebook around 1837. He would go on to sketch 17 such patterns before publishing a "tree of life" diagram in his 1859 *On the Origin of Species*. These are not portrayals of a particular set of species, but rather abstract patterns that are designed to demonstrate the concept of a tree of life. There had been trees that attempted to connect different species before, dating back at least to one produced by the naturalist Jean-Baptiste Lamarck in 1809. But Darwin was looking for a different kind of pattern.

It has been pointed out that his structures are more like a coral with its repeated divisions and convolutions. In fact, Darwin himself wrote in one of his notebooks, "a tree is not a good simile—endless piece of seaweed dividing." A tree has a single trunk from which branches emerge, but the pattern of dividing off of species was far more diverse with split after split after split. The (only) diagram that Darwin included in *On the Origin of Species* is even less like a tree, and more like a spray of decaying particles.

This didn't stop others going for more explicitly treelike patterns. The German naturalist Ernst Haeckel is perhaps the best-known example. Haeckel was fascinated by visual patterns in the structures of organisms. In a book called *Kunstformen der Natur* (Art Forms of Nature), he produced many beautiful illustrations of species variants, their visible resemblances and differences. However, in trying to tie the species together, Haeckel missed Darwin's point and stuck to a pattern that became more and more treelike, culminating in a literal portrayal of a tree of life with a vast trunk.

Haeckel's illustrations were unmatched as works of art, but it was Darwin's not-tree tree that would come to be the dominant pattern and the one that would drive us toward the modern cladogram.

Chaetopoda—a class of segmented worms—is one of the many beautiful illustrations from Ernst Haeckel's book *Kunstformen der Natur* (1899).

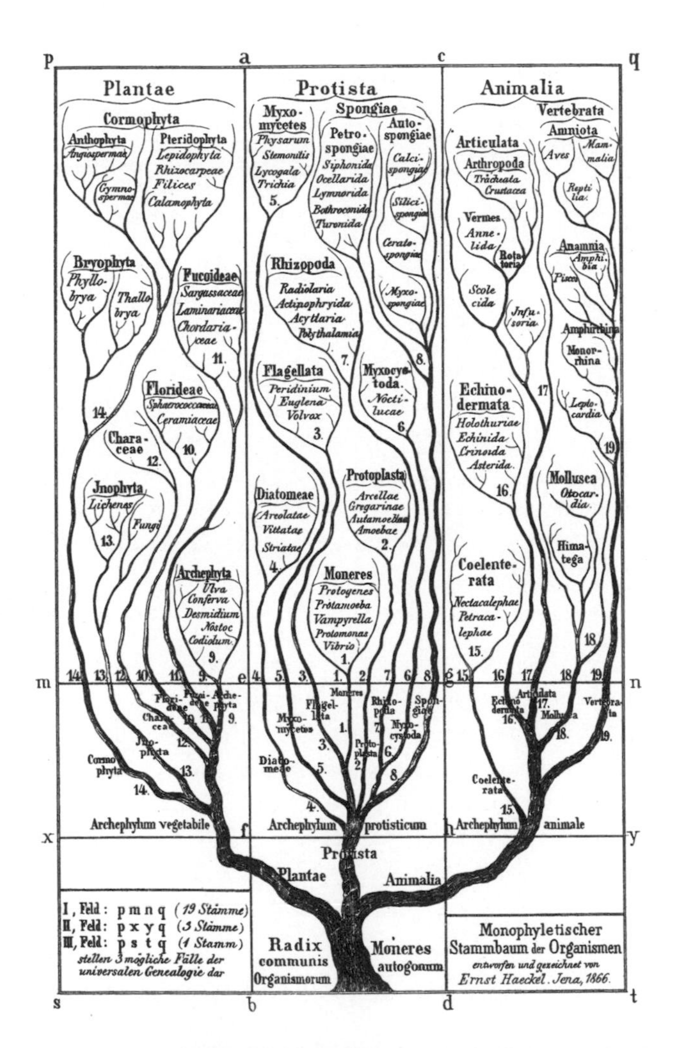

Tree of life by Ernst Haeckel, taken from his *Generelle Morphologie der Organismen* (1866).

FROM GENES AND MORPHOLOGY TO TREE PATTERNS

Like Darwin's original tree, a cladogram shows a relationship between species and their common ancestors, but unlike some of the patterns we have seen before in this book, there is no standard layout for a cladogram. They can be linear or circular. They can (and do) head in whatever direction suits the scientist drawing them. The only underlying certainty is that, like Harry Beck's classic 1931 London Tube map (and those that have followed it), the cladogram is all about relationships between different species (the stations in the Tube map example). Just as the Tube map pattern tells us nothing about distance between stations, the cladogram shows what order the species split off from a common ancestor but nothing about how close or distant those splits are. The underlying logic of cladograms and other related patterns is known by the clumsy term "phylogenetics," roughly meaning the source of a race or tribe.

There is a wide range of other ways to display phylogenetic information using a visual pattern. In many early trees, the "nodes" of the diagram—the points in the pattern where a new branch splits off—were the ancestors themselves. This made them, in effect, family trees. This kind of structure is fine if you know all the detail of the ancestry you are looking at. But, as we have seen with the diagram featuring humans and the great apes, the power of the cladogram is that we are able to produce the pattern even though we don't know what species was the common ancestor of, say, a chimpanzee and a human. With cladograms it is the pattern that is significant, not the nodes in the network.

There are variants of cladograms such as chronograms, which are to scale, so a longer line indicates a longer timescale between species splitting off from common ancestry, but like the family tree, this approach is not always practical. We simply don't always have the appropriate data. Although it is usually possible to tell how old a particular fossil is, this does not tell us how long the species has existed and, bearing in mind how sparse the fossil record is, it is unlikely that we will ever be able to produce accurate timescales for some species.

As we have seen, Darwin did not know what provided the logic for his multiply splitting patterns. But around a century after the publication of *On the Origin of Species* a new and remarkable natural pattern would be revealed, one which provided the data responsible for every phylogenetic tree that depicts evolutionary relationships: the pattern of the structure of DNA.

Cladogram of species

Circular cladogram by David Hillis showing species with fully sequenced genomes.

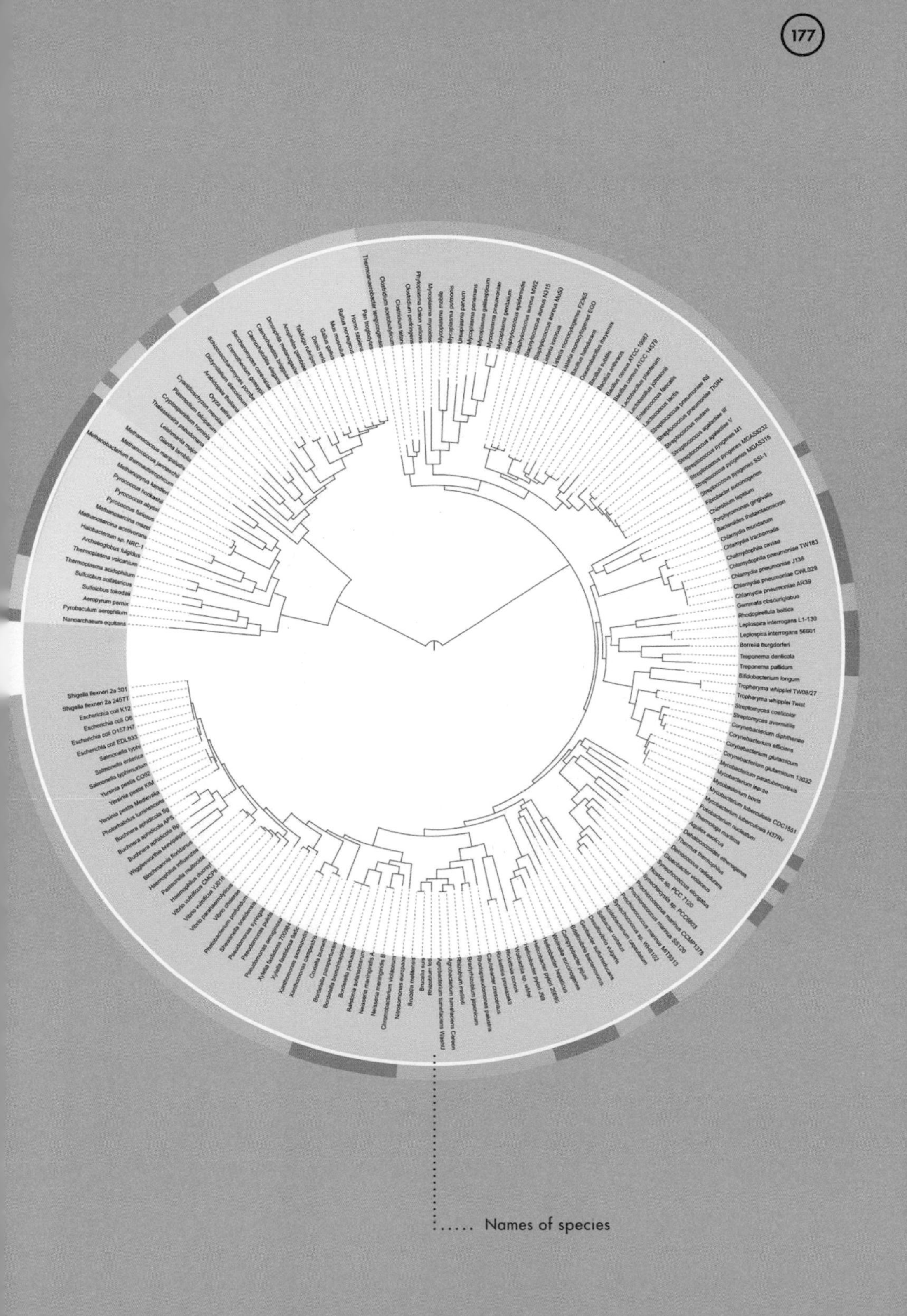
Names of species

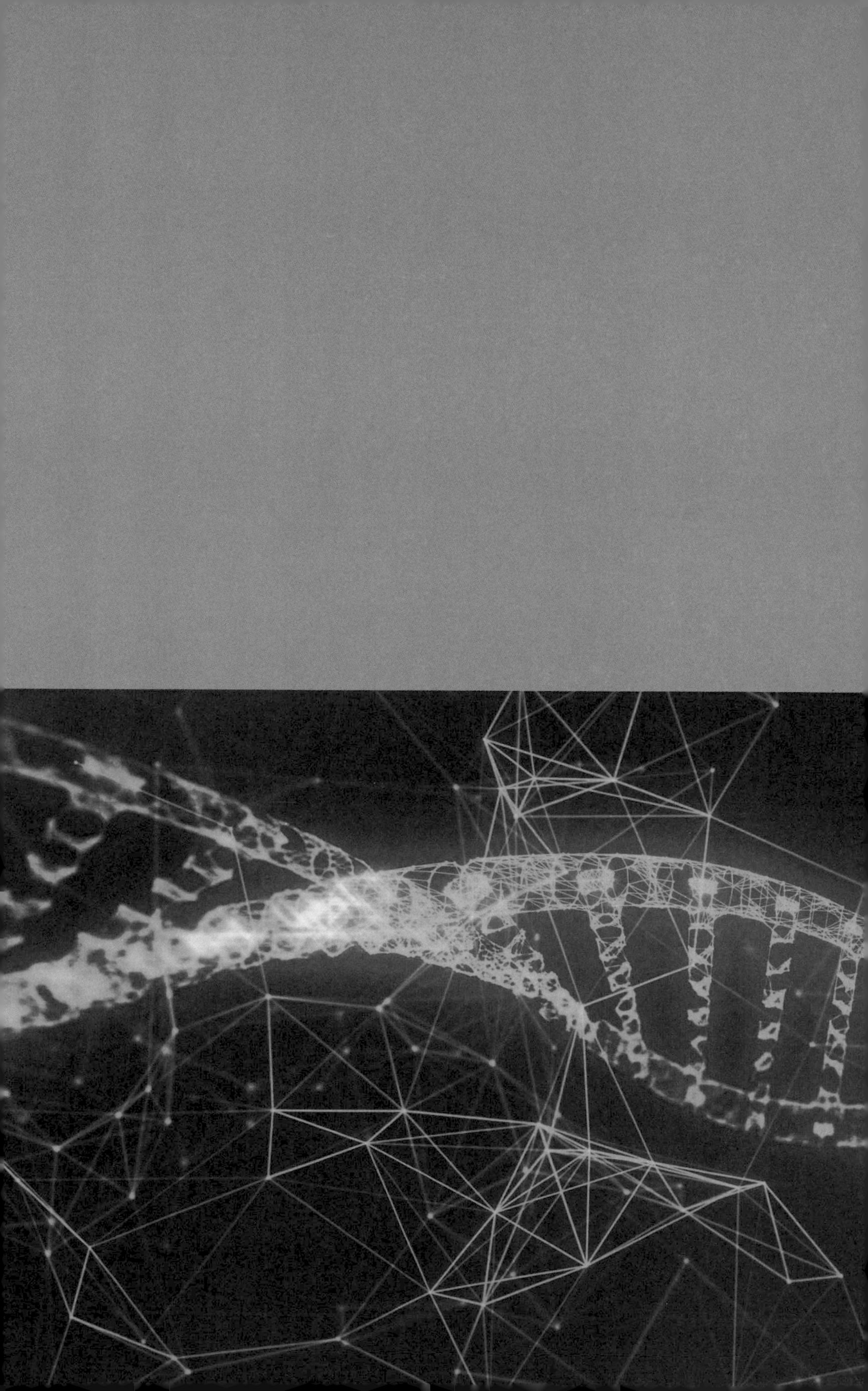

9
DNA DOUBLE HELIX

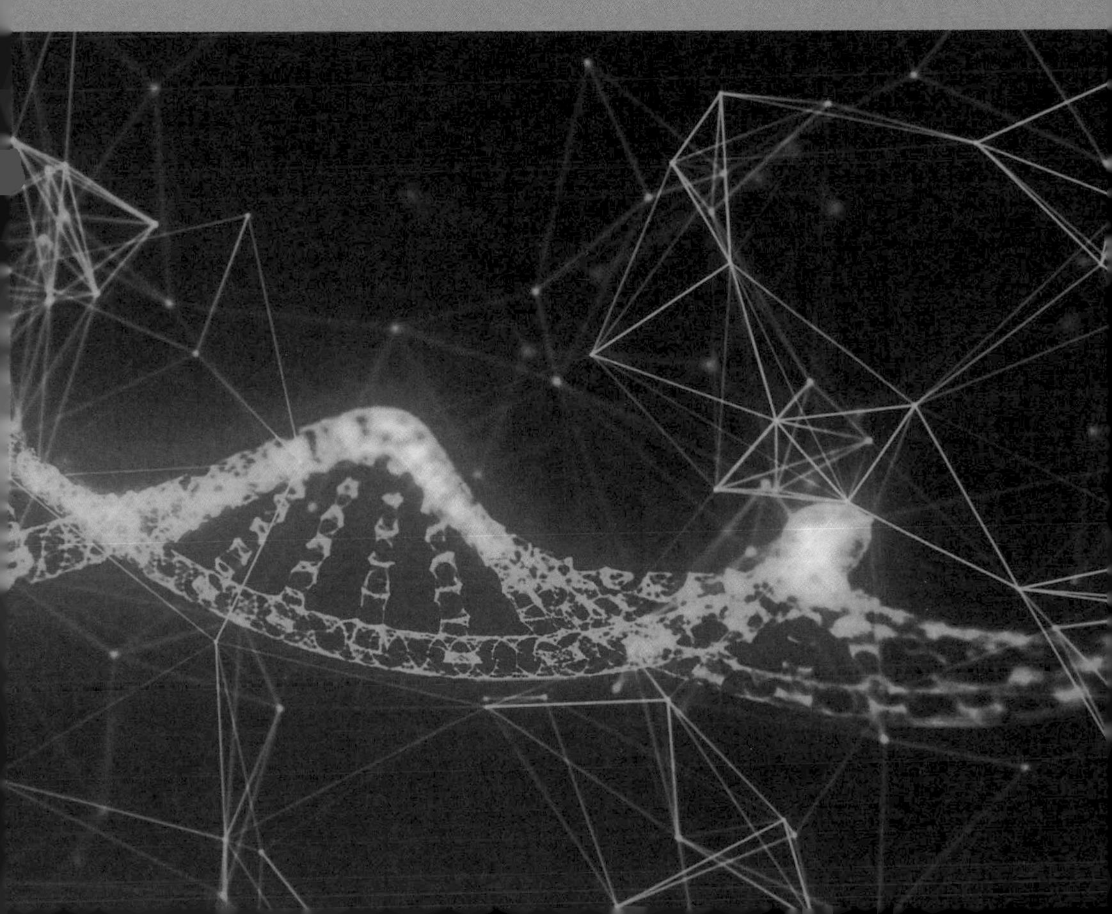

James **Watson** (left) 1928–
Francis **Crick** (right) 1916–2004

THE PATTERN OF DNA

It was in 1953 that Crick and Watson discovered the structure of DNA. The double helix shape of DNA is now so iconic that the molecule can be identified by that simple pattern alone, but it is more than just a decorative shape. The pattern provides the location for the function of DNA: it is a data store, where the pattern is the information it carries and it provides the mechanism for the molecule to reproduce when a cell splits. A molecule of DNA is far more than a collection of genes: that pattern extends beyond the 2 percent of chromosomes that map out protein structures, to include controls that enable DNA to direct living things to develop, reproduce, and survive. This pattern extends across every living creature in the world.

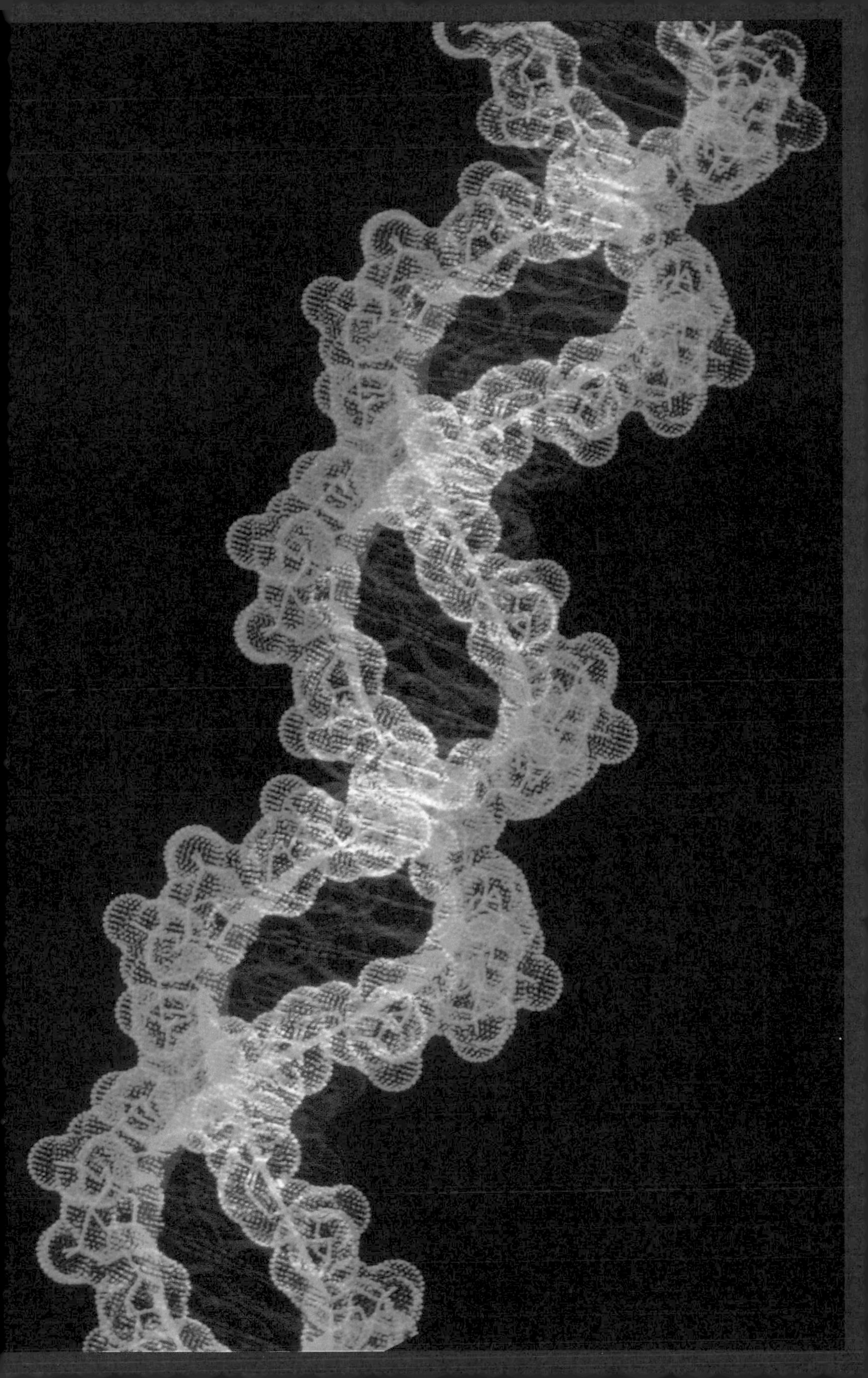

THE INFORMATION MOLECULE

DNA's helical shape is forced on it by the asymmetrical nature of the components that make it up—the twist avoids atoms bumping into each other. But the pattern behind that shape is far more than a matter of physical convenience. This is because DNA's pattern is a store of information, just as much as the memory chip at the heart of a phone or a computer. However, unlike the designed computer chip, DNA is an information store that was developed by the random walk of evolution, which is now responsible for the workings of virtually every living organism.

The letters D N A stand for deoxyribonucleic acid. Of itself, this doesn't sound particularly impressive. It's a reasonably compact name, certainly compared with other organic substances. The official pattern for naming chemical compounds can produce far more complex sounding names, such as the one opposite. For the sanity of chemists, it is usually known as tryptophan synthetase.

This monster full name would be trivial in comparison to that of DNA if we were to name in full any one of the vast arrays of DNA molecules that exist in nature. That's because DNA is not a single compound; instead, it is a framework—a basis for a pattern. Each of your chromosomes, the structures inside the cells that carry your genes, is a single molecule of DNA. The longest of these, human chromosome 1, contains around 10 billion atoms. And these atoms do not occur in a simple repeating pattern, as is the case in many of the large organic compounds. Instead, each of the patterns is distinct from organism to organism and even from individual to individual.

Most of us are familiar with the fact that computers store information in units known as "bits" (short for "binary digit"). Each bit can hold a value of either 0 or 1. A single bit on its own has very limited value, but connect together billions of them and all the digital content we are now familiar with can be stored, ranging from e-books to streaming videos. What's more, inside a computer or phone, that pattern of bits is responsible both for the data and for the code that makes it possible to undertake all the roles that our technology supports. In the same way, DNA makes use of a repeated small-scale unit, but the biological equivalent of a bit can hold any one of four different values. (This makes it quaternary rather than binary, so technically, DNA's fundamental unit should probably be called a quit.)

imagemethionylglutaminylarginyltyrosylglutamylserylleucylphenyl
alanylalanylglutaminylleucyllysylglutamylarginyllysylglutamylglycy
lalanylphenylalanylvalylprolylphenylalanylvalylthreonylleucylglyc
ylaspartylprolylglycylisoleucylglutamylglutaminylserylleucyllysyli
soleucylaspartylthreonylleucylisoleucylglutamylalanylglycylalany
laspartylalanylleucylglutamylleucylglycylisoleucylprolylphenylala
nylserylaspartylprolylleucylalanylaspartylglycylprolylthreonyliso
leucylglutaminylasparaginylalanylthreonylleucylarginylalanylphe
nylalanylalanylalanylglycylvalylthreonylprolylalanylglutaminyl
cysteinylphenylalanylglutamylmethionylleucylalanylleucylisoleuc
ylarginylglutaminyllysylhistidylprolylthreonylisoleucylprolylisoleu
cylglycylleucylleucylmethionyltyrosylalanylasparaginylleucylvalyl
phenylalanylasparaginyllysylglycylisoleucylaspartylglutamylphe
nylalanyltyrosylalanylglutaminylcysteinylglutamyllysylvalylglycyl
valylaspartylserylvalylleucylvalylalanylaspartylvalylprolylvalylglu
taminylglutamylserylalanylprolylphenylalanylarginylglutaminyla
lanylalanylleucylarginylhistidylasparaginylvalylalanylprolylisoleu
cylphenylalanylisoleucylcysteinylprolylprolylaspartylalanylaspar
tylaspartylaspartylleucylleucylarginylglutaminylisoleucylalanylser
yltyrosylglycylarginylglycyltyrosylthreonyltyrosylleucylleucylsery
larginylalanylglycylvalylthreonylglycylalanylglutamylasparaginy
larginylalanylalanylleucylprolylleucylasparaginylhistidylleucylvalyl
alanyllysylleucyllysylglutamyltyrosylasparaginylalanylalanylprolyl
prolylleucylglutaminylglycylphenylalanylglycylisoleucylserylalanyl
prolylaspartylglutaminylvalyllysylalanylalanylisoleucylaspartylal
anylglycylalanylalanylglycylalanylisoleucylserylglycylserylalanyli
soleucylvalyllysylisoleucylisoleucylglutamylglutaminylhistidylaspar
aginylisoleucylglutamylprolylglutamyllysylmethionylleucylalanylal
anylleucyllysylvalylphenylalanylvalylglutaminylprolylmethionyllysyl
alanylalanylthreonylarginylserine

Tryptophan synthetase

The result of spelling out the full name of the compound known as tryptophan synthetase.

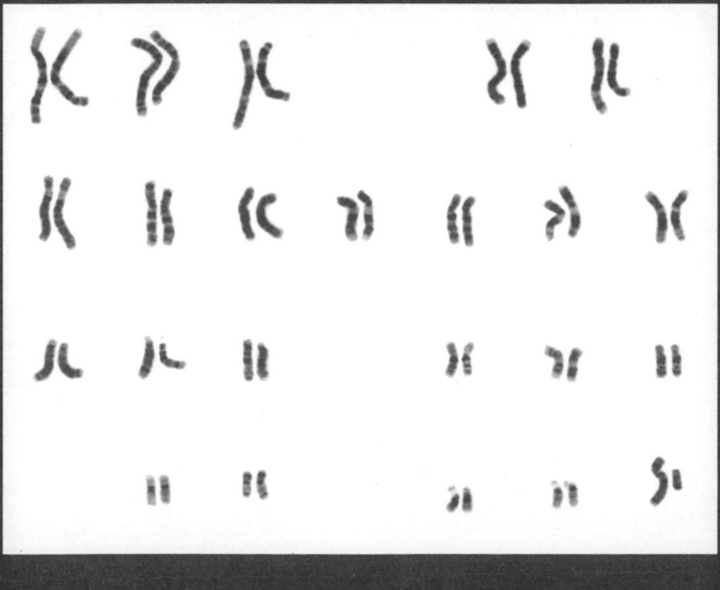

The 23 human chromosome pairs. Each chromosome is a single DNA molecule wrapped around a spindle.

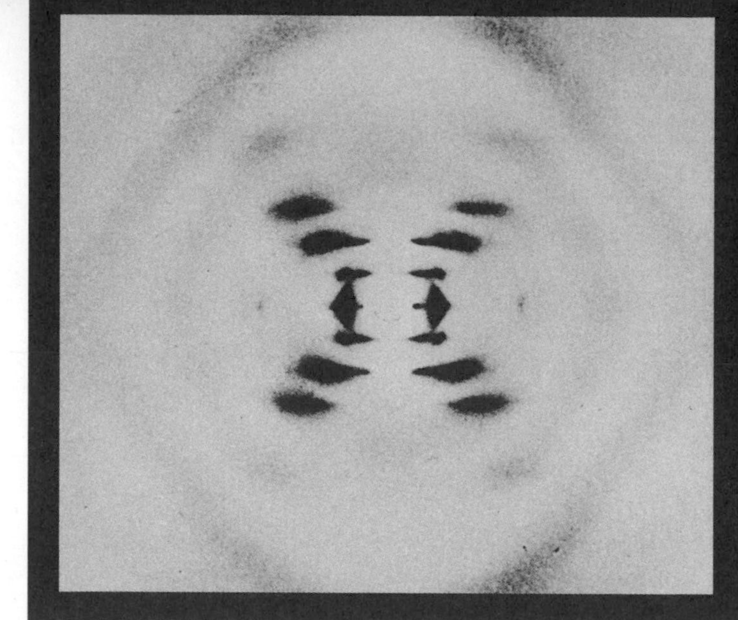

DNA X-ray diffraction image produced by Rosalind Franklin. The result is not a picture of DNA, but rather the effect the structure has on deflecting the X-rays.

THE ESSENTIAL PATTERN

In 1953, Francis Crick and James Watson, working at the Cavendish Laboratory in Cambridge, announced that they had made the discovery of the structure of DNA. Their discovery was based on data produced by Rosalind Franklin, Maurice Wilkins, and Raymond Gosling, who were working at King's College, London using a process known as X-ray diffraction. This technique is a bit like a guessing game where you try to work out what's in a bag without looking into it. X-ray diffraction involves firing X-rays at a molecule and using the way that the X-rays change direction to work out what's happening inside.

The discovery of the structure of DNA is often mistaken for the discovery of DNA itself. But the compound had been known of since 1869, and had been considered to act as a biological store of information in some unknown way for decades. What was crucial here, was discovery of the pattern—the structure that enabled DNA to fulfill this essential role.

PHOSPHATE GROUPS: These form part of a molecule consisting of a phosphorous atom linked to four oxygen atoms. In an organic molecule, such as DNA, phosphate groups have three of the oxygen atoms connected to carbon-based molecules.

The familiar double helix shape of DNA is important, but in the end those elegant, long curved chains of atoms are just a kind of scaffolding. The helices are repeatedly linked sugar molecules, connected by units called phosphate groups. These molecules give the whole compound its name—the sugar is called deoxyribose. When we hear the word "sugar," it is natural to think of the sweet substances we love to eat. Most familiar of these are glucose, fructose, and sucrose, but ribose is another variant, and deoxyribose is ribose with one less oxygen atom than the standard form.

That helix structure is essential to hold DNA together, but the functional parts are found in what would be the rungs, if a piece of DNA were a spiral staircase (see page 186). These rungs are a series of chemical structures known as base pairs, which link the two helices: it is here that we find that essential pattern of information. Each base pair consists of two of four possible compounds called bases: cytosine, guanine, adenine, and thymine. Crucially, the bases always pair off the same way. Cytosine is linked to guanine, and adenine is linked to thymine.

The equivalent of the bits in a computer can be found by reading down the bases that are attached to one helix. Each of these could be any of C, G, A, or T (as the bases are usually named), giving us the four-way quaternary code. But given those values, we know that the base attached to the other helix will always be G, C, T, and A respectively. This might seem to be wasteful redundancy, and natural systems developed by evolution often do feature unnecessary redundancy, but in this case there is an important reason for the repetitive pattern to occur.

DNA is found in most cells in an organism. (Some specialist cells don't have it, for example, red blood cells.) The way that multicell organisms grow and develop is for cells to split, producing two cells from one original. Even single-celled organisms need to split to produce offspring. Each of the cells produced in the splitting process needs its complement of DNA. Therefore, in the mechanism, DNA is first unraveled if necessary (in the kind of complex cell used by all animals and plants it spends most of its time wrapped around molecular spindles called histones) and then it is unzipped down the middle.

At this point, we have two, mirror image copies of the DNA pattern. Molecular machinery units, tiny functional structures built up from sets of chemical compounds, then reconstruct the other half of the DNA molecule. Because the pairing of C with G and A with T is always the same, this can be done, resulting in two copies of the original data. In principle, this process is perfect, although in practice there are occasional errors in the copying process, which is one of the ways that mutations are introduced.

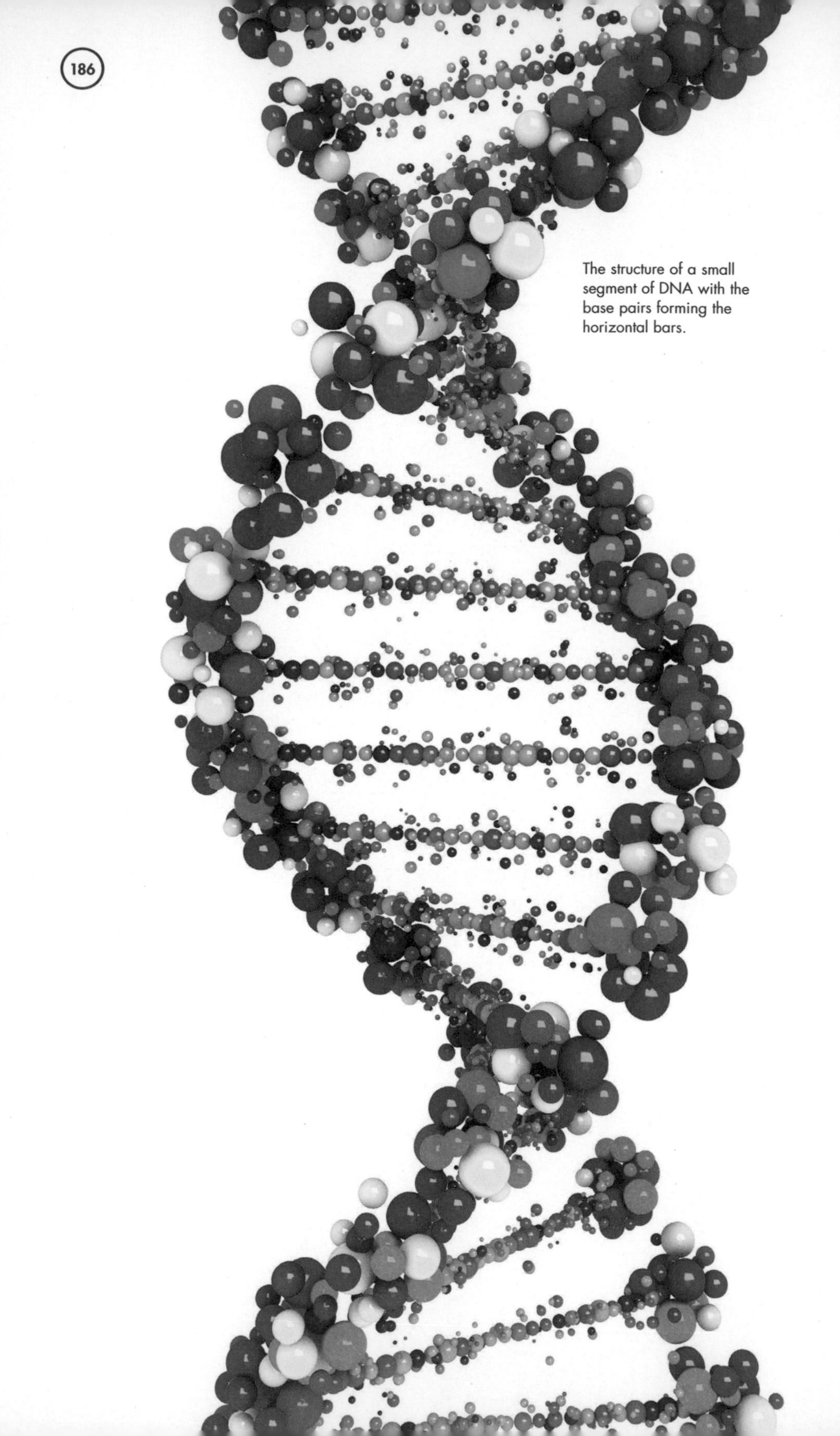

The structure of a small segment of DNA with the base pairs forming the horizontal bars.

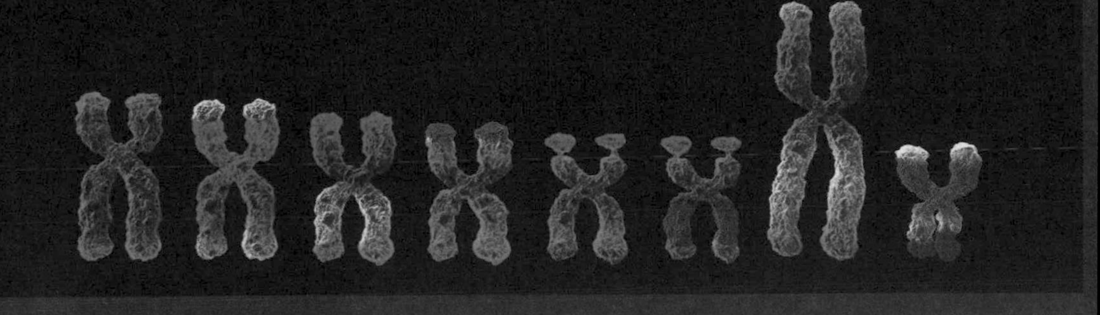

Human male chromosomes (colors artificial), showing the dramatic difference in size between the final two chromosomes X and Y.

THE CHROMOSOME DATA STORE

Most of the cells in a human body contain 46 structures known as chromosomes. Each of these is a single molecule of DNA. These molecules are so large that if the chromosomes from a single human cell were laid out end to end, they would be 6 to 10 feet (2 to 3 meters) long. There are so many cells in your body that your entire collection of DNA would be about twice the diameter of the solar system. The first 44 of those pieces of DNA in a cell come in matching pairs, but chromosome 23, the so-called sex chromosome, has two potential variants. Females have the same broad pattern in both chromosomes, a variant known as the X chromosome, while males have one X chromosome accompanied by the significantly smaller Y version.

Although we all have two of each of the chromosomes (except in males' chromosome 23), the pairs are not identical. One chromosome comes from each of a person's parents. The basic structure is the same, but the detailed pattern of information inside the DNA will contain many small differences, many inherited, others mutations that provide further variation, enabling evolutionary changes to take place. This adds extra variation, rather than the chromosome from one parent being a copy of a single original, each is a mix and match from both of that parent's chromosomes.

A surprisingly large part of the pattern in the chromosomes is common between species. We share about 96 percent of our genes with chimpanzees and as much as 60 percent with bananas. However, the number and size of chromosomes varies hugely from species to species. Prokaryotes, the simpler, single-celled organisms that don't have a nucleus in their cell, such as bacteria, usually only have a single, circular chromosome. Eukaryotes, organisms with a central structure called a nucleus in their more complex cell or cells, have a number of linear chromosomes of varying sizes.

Humans are very much around the middle of the pattern of chromosome numbers and sizes, and there is no correspondence between the apparent complexity of the organism and these values. Some ants, for example, only have two chromosomes to our 46, while a cat has 38. But snails have 54, dogs 78, hedgehogs 90, and kingfishers as many as 132. Even these are dwarfed by some of the plants; there are ferns, for example, with well over 1,000 chromosomes, although it is often the case that the plants have several variants of each chromosome.

The lungfish has a remarkable 130 Gbp genome.

The kingfisher has nearly three times as many chromosomes as humans, with 132 to our 46.

Similarly, the amount of total information in the chromosome package does not reflect the complexity of the organism to which it belongs. In a computer, we refer to the size of the memory in megabytes or gigabytes (where a byte is eight bits). The size of chromosomes is measured using the number of base pairs, usually given in thousands (Kbp), millions (Mbp), or billions of base pairs (Gbp). Aside from viruses, which have very small amounts of genetic material, the sizes range from bacteria, which can have anything from 100 Kbp to 10 Mbp, through insects and many plants in the 100 Mbp to 500 Mbp range. Animals are often in the midrange from around 1 Gbp to 4 Gbp; humans are situated at the 3 Gbp mark. But plants continue on up, reaching all the way to a japonica with around 150 Gbp.

“ THE BEGINNINGS OF LIFE ARE CLOSELY ASSOCIATED WITH THE INTERACTIONS OF PROTEINS AND NUCLEIC ACIDS. FLORENCE BELL

CRACKING THE CODE

It is one thing to know the structure of DNA and the size of these remarkable molecules, but it is another to be able to read off the code and to be able to understand what the pattern means. If you were to look at the pattern of 0s and 1s in your computer or phone memory it would be meaningless to you. You need to know what the significance of that pattern is.

Just as an example, take the pattern:

01000010 01010010 01001001 01000001 01001110

Without the knowledge of what the binary pattern indicates, it would be hard to deduce what these numbers are telling us. If, though, we knew that this was the ASCII code used to represent characters in a computer (see below), it would be possible to read off the pattern as representing the word BRIAN.

ASCII CODE: A code for representing letters and other characters with the binary numbers (based only on 0 and 1) used by computers. The basic ASCII code uses eight bits for each letter.

The best-known part of the information encoded in the DNA pattern makes up the genes. These are chunks of the code that is used to describe the structure of other molecules, most often one of a range of vital building blocks of life called proteins. In doing so, genes have important roles in specifying many of the instructions for the construction of an organism. Some will operate together in complex sequences, while others operate independently or in small groups to define simple aspects of the organism, such as hair color or eye color. Variants of genes, often the result of accidental changes known as "mutations," produce the differences between individuals. Each of us carries many such variants—we are all mutants—and over time the accumulation of these changes results in the development of new species as seen in the previous chapter.

Proteins are themselves complex molecules, constructed from building blocks called amino acids. It is here that the genetic code comes into play. There are 20 such amino acids, so a single base pair value is insufficient to identify which acid comes at a particular location in a protein. Again, there is something of a parallel with computing. The 0 or 1 values held in a bit are obviously not enough to specify a particular letter of the alphabet. Initially these were specified by an eight-bit word, allowing for

128 characters in the ASCII code (for technical reasons, only seven of the bits were used). That was fine for a single alphabet—as in the code for BRIAN opposite—but computers now support multiple alphabets and special symbols, using an extension of ASCII called Unicode, where each character is represented by an extendable set of at least 16 bits.

In the case of the genetic code, it is triplet patterns of base pairs, known as codons, that spell out the amino acid options. As each base pair can have one of four values, this provides a total of 64 possible values—more than enough to specify the 20 amino acids. In practice, there is an additional value, because a protein is constructed from multiple amino acids strung together. When the molecular machinery that assembles a protein is reading off a series of codons, it needs an instruction to stop and complete the molecule, so there is also a codon for a "stop" command.

AMINO ACIDS: Organic compounds combined in chains to produce proteins. They contain a nitrogen/hydrogen group and an acid group in the form COOH.

If the DNA code had been designed like ASCII, rather than having evolved, it would have made sense to have one code for each amino acid or control command, leaving over 40 free for other possible uses. However, as we have seen, nature isn't designed, so each of the amino acids and the stop command have several codons all producing the same outcome. Every possible combination of the 64 is used up.

ALMOST ALL ASPECTS OF LIFE ARE ENGINEERED AT THE MOLECULAR LEVEL, AND WITHOUT UNDERSTANDING MOLECULES WE CAN ONLY HAVE A VERY SKETCHY UNDERSTANDING OF LIFE ITSELF.

FRANCIS CRICK

BEYOND PROTEINS

If the only role of DNA were to act as a map for proteins, it would not need to be bigger than the set of genes for that species. The number of genes an organism has varies hugely; humans have around 25,000–30,000 genes coding for proteins, while rice has more than 45,000 such genes. In practice, this is only a tiny part of the total chromosomes available. A relatively small part of the extra space is taken up by a number of genes that are not used to specify the structure of proteins, but rather of RNA.

RNA: Ribonucleic acid is similar to DNA as an information store, but only has a single strand and uses one different base—uracil—instead of thymine.

RNA can be described as a sort of cut-down DNA, which only has a single strand, rather than the dual helix, with a subtle variant in the bases, replacing thymine with uracil. RNA has a number of roles. In some viruses, for example, it is the sole source of genetic material. One of its main functions in the vast majority of organisms is to take the information from DNA and turn it into a template for constructing a protein. In the process, the DNA is partially unzipped and a strip of RNA is created to reflect the section containing the gene. However, simply copying the pattern is still insufficient to do the job, because a gene does not consist of all the codons required to produce a protein in a neat, unbroken sequence.

Editing RNA

In producing messenger RNA, the introns (shown as gray in the pre-mRNA) are edited out to produce a continuous protein coding section of code.

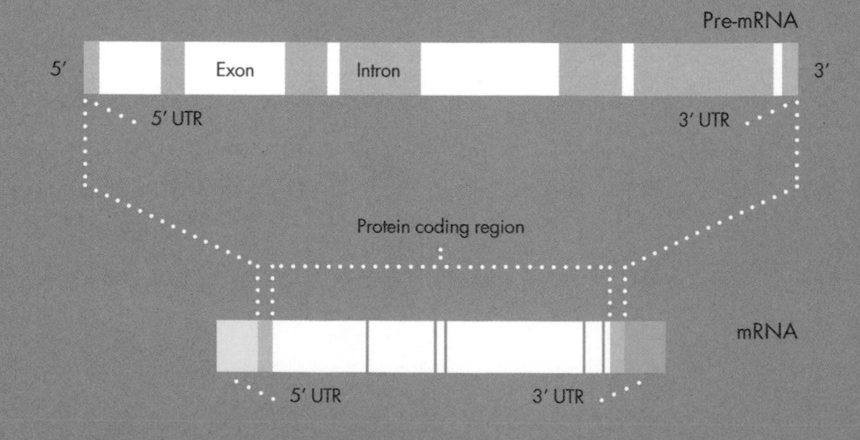

Rice plants have far more genes than human beings, despite being a significantly simpler organism.

The information in DNA was not put together in a structured, directed fashion, but by the random wanderings of evolution. As a result, there are chunks of DNA between the codons known as introns. After the first piece of RNA is constructed (in a process known as transcription), it then has to be edited to cut out the introns, so that there is a correct sequence of codons to assemble the protein.

As well as the "messenger RNA" that is transcribed from DNA, there are other variants of this versatile molecule used as enzymes alongside proteins. There are also short lengths of RNA used to link messenger RNA to amino acids; these types of linking RNA have their own genes over and above the "coding genes" that specify proteins.

This still only accounts for a few percent of the whole of the pattern encoded in a chromosome. The rest has traditionally been referred to as "junk" DNA, reflecting the way that evolution piles in different functions without tidying up the structure. Some of the extra DNA is repetition as a result of copying errors. Other parts may have come from other organisms, such as viruses, which are able to modify DNA structure. But it has relatively recently been realized that other stretches of DNA have very important functions.

A representation of methylation where the two bright points show the location of methyl groups disabling bases.

THE EPIGENETIC CODE

The extra functioning DNA is referred to as epigenetic—over and above the genes. This is still material that is inherited, but if we look at our analogy with computers, where the genetic code is like data, the epigenetic code is more like part of the computer program. It provides mechanisms for switching genes on and off, for example, so they can operate at certain times in an organism's lifetime, but not be functioning at others.

One of the most common epigenetic mechanisms is called DNA methylation. In chemistry, a "methyl group" is a simple structure of a carbon atom bound to three hydrogen atoms. This leaves the carbon with one more free bond that can link it to something else. When a particular base on the DNA strand is methylated, a methyl group is stuck onto it using this spare bond. As a result, although the sequence of bases is still the same, the methyl group effectively gets in the way, disabling the gene. This function can also be provided by so-called "repressor proteins."

One fascinating aspect of epigenetics is that not only is it influenced by the environment of an organism, but also the epigenetic pattern can be passed on to that individual's offspring. This brings back into evolution an aspect of biological history that had been dismissed as a scientific error. In trying to explain how traits were passed on from generation to generation, the French biologist Jean-Baptiste Lamarck suggested that organisms developed traits as they interacted with their environment, and these traits were then passed on to their offspring.

So, for example, Lamarck suggested that by constantly stretching up to eat the tender, high leaves off trees, the ancestors of giraffes would stretch their necks, making them a little longer. They would then have children that would be born with slightly longer necks; the offspring would then stretch them even further, and so on, leading to a different driver for evolution. This idea was carried to the extreme with the suggestion that the experiences a mother had while a child was in her womb could alter the developing child. So, for example, the noted Victorian Joseph Merrick, who was known as the "Elephant Man," had a disorder producing a distorted skeletal structure. He was said to have developed this condition when his mother was scared by an elephant while Merrick was in the womb.

The idea of Lamarckism would be dismissed and eventually ridiculed as naïve when the concept of genetic inheritance became known as the mechanism behind evolution through natural selection. However, epigenetics has revealed that it is possible for some environmental impacts on the epigenetic factors of a parent's DNA to be passed on to the next generation, producing a limited kind of Lamarckian inheritance. For example, following the Dutch famine during the Second World War, children were born smaller and suffered more from some diseases in adulthood for at least two generations.

Lamarck suggested that giraffes had stretched their necks reaching for high leaves and had passed this change on to their offspring, which over generations would produce longer and longer necks.

NO MORE BLUEPRINT

Epigenetics is the solution to the puzzle that the number of available genes does not provide enough data to make for a full "blueprint" of how to construct, say, a human being. The epigenetic data is an additional part of the pattern that shows how, when, and where to make use of genetic information. If we think of the construction of an organism as being like the output of an automated factory, the genes are similar to the machinery available to produce the product, but epigenetics provides the control program to switch different parts on and off and route materials around the factory.

All eukaryotic organisms—including all animals and plants—start as a single cell, which repeatedly splits, doubling its complexity each time. If the cells were only able to split, all that would produce would be an undifferentiated blob of identical cells. However, what happens is a process called cellular differentiation. The initial multipurpose cells are switched to have a variety of functions, producing very different types of cell from, for example, the elongated neuron to muscle cells. It is epigenetics that enables this switching to take place.

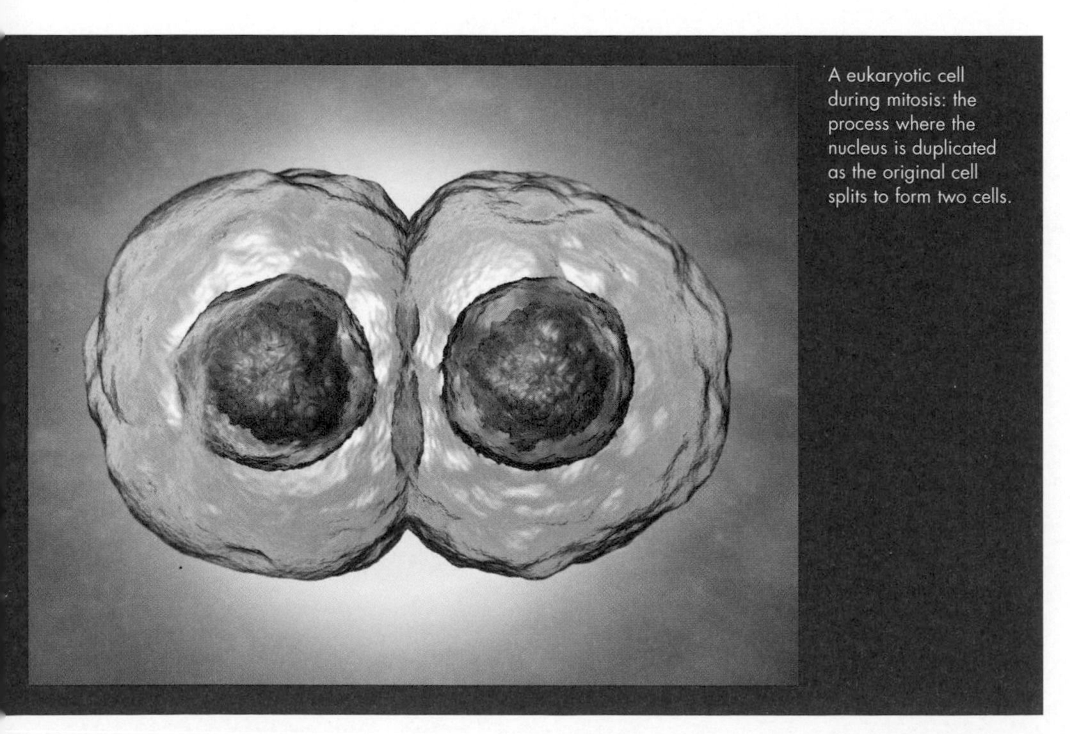

A eukaryotic cell during mitosis: the process where the nucleus is duplicated as the original cell splits to form two cells.

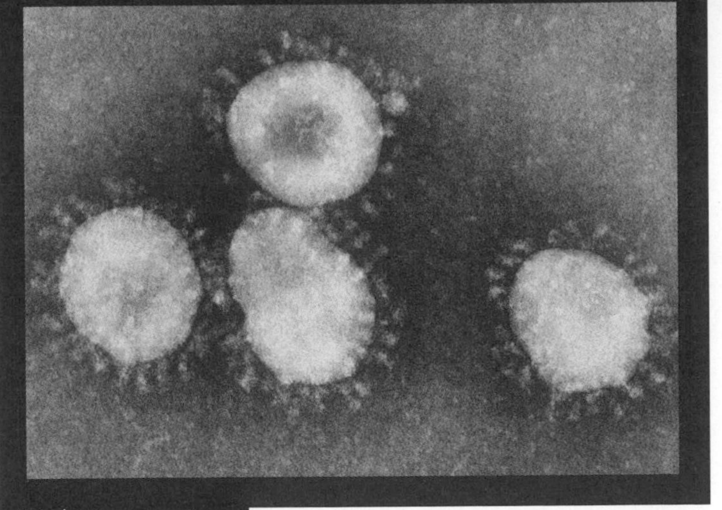

Actual avian coronavirus particles, taken with a transmission electron microscope.

At the same time, as well as providing different types of cell, a developing organism has to go from collections of cells to forming the physical structures that make it up, such as organs. In this process, some genes produce compounds called morphogens, which act as a kind of three-dimensional blueprint and turn on genes. In effect, although created by genes, morphogens go beyond the simple genetic structure to provide an epigenetic pattern that enables the final organism to form.

Lying beneath all of this, and responsible for the existence of almost all the living things we know, from bacteria to animals and plants, is the guiding pattern of DNA. Although a few organisms make do with RNA alone, DNA is universal elsewhere. The data that the DNA stores is different in every organism, but its underlying mechanism and coding is identical, giving every indication that all living things now in existence had a single common ancestor in the past; it appears that all current life on Earth sprang from a single origin.

We don't yet know if there is any other life in the universe, other than that found on Earth. Although science fiction likes to portray aliens, we have no scientific evidence that extraterrestrial life exists. If it does, it would be surprising if it worked in exactly the same way as we do, but there would need to be some equivalent to DNA—a versatile chemical information store that could be passed from generation to generation (unless the alien were a single entity with no offspring). DNA provides a vital role in sustaining life in changing environments, but our final pattern is even more fundamental. Symmetry is arguably the pattern that lies beneath reality itself.

10
SYMMETRIES

Emmy **Noether**
1882–1935

SYMMETRY PATTERNS

Symmetry is a term we associate with art and architecture—and it is important there—but the patterns of symmetry produced by reflection, rotation, translation, and more are central to many aspects of nature, from the biological symmetries of animal bodies to the underlying symmetries of physics that determine the numbers of types of particle, and even underpin fundamental laws, such as the conservation of energy. Scientists believe that symmetry is embedded in the patterns of the universe, shaping much of the way that the natural laws are established at a deep level. The pioneering German mathematician Emmy Noether would prove mathematically that the conservation laws, physical laws that are responsible for much of the everyday behavior of the world around us, were the result of symmetry in nature. Symmetry is far more than a visually attractive layout: it is the pattern that shapes the heart of reality.

MIRROR, MIRROR, ON THE WALL

Symmetry (literally meaning "measured in a like way") is one of those words that we employ a lot without thinking too much about what it really means. In general usage, it is most commonly applied to what is known as "mirror symmetry." This is not about seeing our own image in a mirror, but rather about patterns that reflected in a mirror are identical to the original. Such symmetry can be produced by dividing an image in two and replacing one half with a mirrored version of the other. Mirror symmetry is a pattern where one half of something is a reflected version of the other half.

CHIRALITY: Objects, such as gloves, that have "handedness" are described as "chiral" scientifically. Chiral molecules have the same chemical formula but a different shape, reacting differently.

Understanding what this means is not helped by the fact that we mostly have an incorrect mental picture of what a mirror does to what we see. Look at yourself in the mirror. The image in front of you is not the same thing that someone else sees looking at you face on. Hold up your left hand and your reflection seems to hold up its right hand. It feels as if the mirror swaps things left and right. But if that really were the case, how does the mirror know to do the reversal left to right but not to swap top and bottom? There is nothing special about the orientation of the mirror: turn it through 90 degrees and it still seems to swap left and right.

You can get a better idea of what is happening by holding up a book (or magazine) with the front cover facing the mirror. Your reflection is holding the magazine with its back cover facing you,

"SYMMETRY IS WHAT WE SEE AT A GLANCE; BASED ON THE FACT THAT THERE IS NO REASON FOR ANY DIFFERENCE. BLAISE PASCAL

Symmetry is often used in art for its pleasing effect, as demonstrated in this mirror-symmetrical image.

although that back cover has on it a flipped version of whatever is on the front cover of your book. What this shows us is that the mirror actually turns the image inside out, as if the items on view were rubber shells. When you raise your left hand, you interpret what you see in the mirror as your reflection's right hand, but, in reality, it's the same hand inverted front to back.

With that in mind, we can see that mirror symmetry (echoing "like measurement" in the definition) involves taking the distance of each point in half of an image from an imagined mirror and reversing it in the other direction. Each point on both halves of the mirror-symmetric image is the same distance as its equivalent point, measured from the line of symmetry where the reflection takes place.

The tiger's camouflage pattern is close to symmetrical down its spine and forms part of its aesthetic appeal.

TYGER TYGER

We find mirror symmetries attractive.

Many of the architectural styles common before the twentieth century, for example, tend to have mirror symmetry. This pattern of attraction originates in the human physique. We find people attractive who have symmetrical faces and bodies. Across centuries and cultures there have been different appearances that have been considered particularly beautiful or handsome. Women, for example, have been considered attractive when slim or fat, clear skinned or tattooed, pale of complexion or tanned, with small noses or large ones, and so on, through an endless and bewildering series of contradictions. But facial symmetry is an unchanging marker of why a man feels that a woman is beautiful, or a woman thinks a man is handsome.

This preference is natural selection at work. The outward signs of many diseases produce a degree of asymmetry—when we see a face that has perfect balance, we assume it reflects health. This enthusiasm for symmetry has been tested using photographs where half faces are mirrored: the symmetrical version is nearly always considered more attractive than the original face, if not as interesting. At the basic biological level, when looking for a mate we are seeking out the ability to breed and looking for lasting health. Symmetry, attractive in both sexes, has historically acted as a marker of health.

BILATERAL SYMMETRY: Around 99 per cent of animals have bilateral symmetry, meaning that their bodies have rough mirror symmetry on a line down the middle. Internally this symmetry tends to break down, for example, the position of the human heart.

The same preference is noted among chickens, which also prefer a symmetrical member of the opposite sex, and for that matter in our appreciation of the looks of other species. It's not for nothing that William Blake wrote in his 1794 poem "The Tyger":

Tyger Tyger, burning bright,
In the forests of the night;
What immortal hand or eye,
Could frame thy fearful symmetry?

Mirror symmetry is visually powerful, but it is only a tiny part of the wider scope of the patterns of symmetry.

"SYMMETRY REPRESENTS ORDER, AND WE CRAVE ORDER IN THIS STRANGE UNIVERSE WE FIND OURSELVES IN.

ALAN LIGHTMAN

Though octagonal, this inner well of a building in Germany only exhibits two-fold rotational symmetry.

TWISTING THE LOOKING GLASS

The next most obvious pattern of symmetry involves rotation. Something has rotational symmetry if it is indistinguishable from its original form after rotating it by a certain amount. Imagine, for example, a simple square. Rotate the pattern by 90 degrees—or any multiple of 90 degrees—and it appears to be unchanged. But rotate it by 45 degrees and it looks entirely different. A square has four-fold rotational symmetry. This means that it is symmetrical when turned by a quarter of the full rotation of 360 degrees.

Compare this with a rectangle that isn't square. Turn this through 90 degrees and there is a difference. To achieve an indistinguishable image, we need to turn the rectangle through 180 degrees—half a full rotation—so it has two-fold symmetry. Equilateral triangles (and the ancient symbol of the triskelion, found in Bronze age carvings and the three-legged symbol used in heraldry) have three-fold rotational symmetry (see opposite). We can go on up through polygons with multiple sides until we reach a special case for rotation, unlike any other: the circle.

A circle has the ultimate in rotational symmetry; you can turn it by any amount and it always looks the same. You could describe this as being infinite-fold symmetry. This is not only a matter of abstract visual or mathematical interest. The circle gives us the first hint of why symmetry is not just of visual interest, but can also be important in the world around us. It is because of the circle's symmetry that the best wheels are round. The infinite-fold of symmetry ensures that as the wheel turns, a part of it will always be in contact with ground that is flat.

This may seem obvious, but there is more to the relationship of symmetry and the wheel than meets the eye. The symmetry of circular wheels is not ideal on all terrains, for example. You could imagine a regular, jagged terrain where a wheel with a number of points would be more effective. However, the chance of having the right wheel shape to match an actual terrain is remote, so a circular wheel is still usually the best option, especially when we have modern, relatively flat road surfaces.

Now that we have a second type of symmetry, we can see that the patterns of mirror symmetry and rotational symmetry overlap, but are not identical. A square and a rectangle are both mirror symmetrical in two directions, with the mirror passing through the center of the shape and parallel to two of the sides. An equilateral triangle has mirror symmetry on three lines, each passing through one vertex (the points on a geometric shape) and the midpoint of the opposite side. And a circle has mirror symmetry on any line through the center. But other rotationally symmetrical shapes, such as the triskelion have no mirror symmetry.

The three-fold rotational symmetry of the triskelion was used in Bronze Age carvings and appears, often in the form of three legs, in heraldry.

These examples all reflect (pun intended) the application of symmetry to two-dimensional shapes, but the same concepts of symmetry can also be applied to three-dimensional objects. Just as we can consider mirror symmetry in a three-dimensional building, we can have rotational symmetry in an appropriate three-dimensional object, from the trivial case of a sphere, through cubes, to more complex structures.

LOST IN TRANSLATION

Perhaps the least obvious example of spatial symmetry (there are other kinds, such as symmetry in time) is translational symmetry. This is the symmetry of sideways movement—how something looks when it has been moved sideways (as opposed to being rotated), compared with the way it looks before being moved.

Trivially, every object, whatever its shape, has translational symmetry with an identical object in a different location. This is because you can move the first object a certain distance and direction so that the two objects are overlaid entirely. Where things become more interesting (and practically useful) is when we consider the translational symmetry of a repeating pattern.

Think, for example, of the pattern of a neatly constructed brick wall. We can slide the wall along by the length of one brick and both the shifted wall and the original wall will appear visually identical, so they appear to have translational symmetry under the operation of moving by one brick's length in that direction. Or do they?

If we imagine a pair of real identical brick walls, one in front of the other, and moved the front wall along by one brick's length, most of the bricks would line up, but there would be a section of each wall that was overhanging at the ends of the run of bricks where the symmetry broke down. The symmetry would not be complete. As a result, strictly speaking, translational symmetry of a pattern only holds up if the pattern is infinite in the direction that the movement is made.

In the real world, we don't know of any physical object that is infinite. As we have seen when looking at the number line, infinity is a powerful mathematical tool, but it can't usually be applied directly to real world objects. Therefore, for real world translational symmetry, we essentially look the other way and ignore the edges, only considering what happens in the remainder of the pattern. As long as it works for the bulk of the pattern, we can still consider it to be meaningful.

PHYSICAL INFINITY:
Infinity is a powerful mathematical concept, but we don't know if anything physical is infinite. The part of the universe we can see is around 90 billion light years across, but we don't know if it continues forever.

A regularly laid brick wall has translation symmetry, as moving the entire wall along by the length of one brick produces the same pattern.

We won't look at them in any detail, but it's worth noting that other kinds of spatial symmetry can be produced by combining different symmetry elements. So, for example, there is spiral symmetry, which combines rotation and translation, and glide reflection, which is a combination of a reflection and a translation.

"TO A PHYSICIST, BEAUTY MEANS SYMMETRY AND SIMPLICITY.

MICHIO KAKU

PASTING AND GROUTING SYMMETRY

In art and other situations where we are dealing with flat surfaces, many of the applications of symmetry can be brought down to the symmetry of wallpaper and of tiles. The patterns of a wallpaper typically have two symmetries: the symmetry that runs along the roll of wallpaper, where the patterns eventually repeat, and the sideways symmetry across a wall. The symmetry along the roll is a simple translational symmetry. If we take any point on the roll, we can find a matching point simply by moving in a straight line, heading straight down the roll. To get to the same point on an adjacent piece of wallpaper, we always have to move sideways from piece to piece, but may also have to move up or down, as the adjacent piece of paper may well have been shifted in the vertical direction. Those two translations can be combined as a single diagonal move.

An Islamic tiling pattern, showing a complex mix of symmetries.

Depending on the symmetries of the elements in the wallpaper pattern, and their location, wallpaper symmetry may also involve rotations or reflections. There are a total of 17 different types of symmetry structures available for wallpaper patterns. Of course, moving away from literal wallpaper, there is no need to be limited to two dimensions; the symmetry options become even richer when considering three-dimensional applications of such patterns, which are particularly apt when considering the symmetry of crystals (see pages 212–213).

Tiling is a rather different approach to the patterns of symmetry. Pleasing patterns in tiles can be created by using all kinds of symmetry. Often, the tiling, like a wallpaper pattern, features a repeat that gives one set of symmetry patterns across the whole, however, there can also be more localized symmetries in tiles that aren't continued across the whole wall or floor. For example, many of the complex tiled patterns used in the tiling of Islamic architecture make use of a whole range of designs that are locally symmetrical, but which aren't symmetrical across the whole tiled surface as the designs shift and interlock.

It might seem inevitable that when tiling with a small number of shapes that the outcome would be that after a while the pattern would be forced to repeat. However, the English mathematician Roger Penrose showed mathematically that with just a few elements it was possible to tile a surface in such a way that there was never complete repetition.

In his twenties, Roger Penrose had explored intriguing visual aspects related to symmetry with his psychiatrist father Lionel. Together, they devised two classic mind-bending patterns: the Penrose triangle and the Penrose staircase. However, Penrose would go further with Penrose tiling. Here, just two shapes can be sufficient to produce a tiled pattern that will continue forever, but will never repeat. As with Islamic tiling, Penrose tiling (for example the image on pages 198–199) can have local symmetric patterns, but across the design as a whole, symmetry continues to be broken, never leading to a pattern of continued repetition.

THE CRYSTAL WORLD

The most obvious bridge between visual patterns and the physics of the real world around us lies in the structure of crystals. The term "crystal" does not just refer to a transparent gemstone, it is any substance that is made up of a regular, repeating structural pattern of atoms. The symmetry considerations we have already seen applying to wallpaper and tiles also apply to the layers of crystal structures, which often exhibit the different types of symmetry pattern. How these symmetries occur can have a significant impact on a substance's chemical and physical properties. For example, the remarkable material known as graphene, a one-atom thick layer of carbon with the crystalline structure of graphite, owes its great strength and electrical conductivity to its symmetry pattern.

Another, perhaps more familiar, example of the power of crystalline symmetry is in the snowflake. This remarkable six-sided structure was first discovered by the Swedish cleric Olaus Magnus back in 1555, although the full wonder of variety of the shapes was only made clear with the introduction of microscopes in the early seventeenth century. Even then, the beauty of snowflakes was not widely appreciated until the American meteorologist Wilson Bentley started capturing images of snowflakes with early photographic technology and techniques in 1885.

The molecular structure of a crystal, such as diamond, exhibits symmetry in its three-dimensional repeating patterns.

Six-fold symmetry in some of the many and varied snowflake photographs from Wilson Bentley's *Snow Crystals*.

Bentley had a lifelong fascination with snowflakes and produced a book of microscope photographs of them near the end of his life in 1931. Entitled *Snow Crystals*, this classic work contained a remarkable 2,000 photographs. It was Bentley who first made the often-quoted observation, based on his life work, that "no two snowflakes are alike." There is no scientific foundation for the idea that each snowflake has a unique pattern, and it is easy enough to find identical flakes in the simpler shapes. But it is certainly true that there is a vast variety of snowflake forms.

The traditional, delicate snowflake with six arms, the kind we duplicate in Christmas decorations, (called "dendritic" or treelike) grows when temperatures are particularly low. But when it is warmer, with the air closer to freezing point, snowflakes tend to form simpler six-sided platelike crystals. The apparent unique nature of snowflake shapes is because their growth is governed by chaos, the mathematical concept we met in the weather patterns section, where very small changes in initial conditions can result in very large differences in the outcome.

The six-sided aspect that gives the snowflake both six-fold rotational and mirror symmetry reflects the molecular form of water, which consists of an oxygen atom with two hydrogen atoms attached, each at about a 104.5 degree angle from each other. A combination of this molecular shape and hydrogen bonding—the electrical attraction between the relatively positive hydrogen and relatively negative oxygen atoms—means that water naturally forms crystals in a six-sided lattice. As these molecular-scale crystals grow, that six-sided form extends into the exotic six-armed snowflake patterns that are so familiar.

THE SYMMETRY OF THE MATHEMATICAL MIND

Because we have such a strong visual idea of what symmetry means, it's important to stretch the mind beyond this, as the power of symmetry really comes into play when it is approached mathematically. Here, symmetry is about algebra—specifically about how one pattern of numbers is translated into another. There is a symmetry when the transformation leaves some of the values unchanged. It is this approach to symmetry that lies beneath a surprising amount of modern physics.

In purely mathematical terms, symmetry is often about dealing with matrices, which are structures of numbers in a table form that have their own mathematical methods for the way they multiply and combine. Matrices are valuable in describing many aspects of physics, and the transformations of one matrix into another can often have symmetry implications. This comes through particularly strongly in the use of symmetry groups.

MATRIX MATHEMATICS: One of the most significant aspects of the mathematical manipulation of matrices is that, unlike ordinary numbers, multiplication is not commutative. These means that A × B is different from B × A.

Group theory is a branch of mathematics that extends the set theory that we met in Chapter 7 on number lines. A group is a set where it is possible to take any two members of the set and manipulate them in a way that produces a third member (subject to a few technical considerations). A simple example of a group is the set of integers. The operation of addition will always translate two integers into a third, making integers a group.

A group has the natural feel of the patterns of mathematical symmetry, but when an object has any kind of symmetry, there will be an associated group called its symmetry group (for example, see opposite). This is represented by a set of matrices that show how the object can be transformed without changing any one aspect of it. Symmetry groups have their own obscure notation involving letters and numbers. If we look at the different ways a sphere can be rotated, its symmetry group is called SU(3), standing for "special unitary group of degree 3."

A simple application of symmetry groups is found in studying one of the most popular physical pattern-related puzzles of all time: the Rubik's cube. When a player rotates a segment of the cube, they are producing a rotation, and the number of possible ways the cube can be arranged is down to the symmetry group of the transformations possible under rotation of the faces. The size of the symmetry group is approximately 519,024,039,293,878,272,000, which corresponds to the number of ways the cube can be arranged. This is not back-of-an-envelope mathematics.

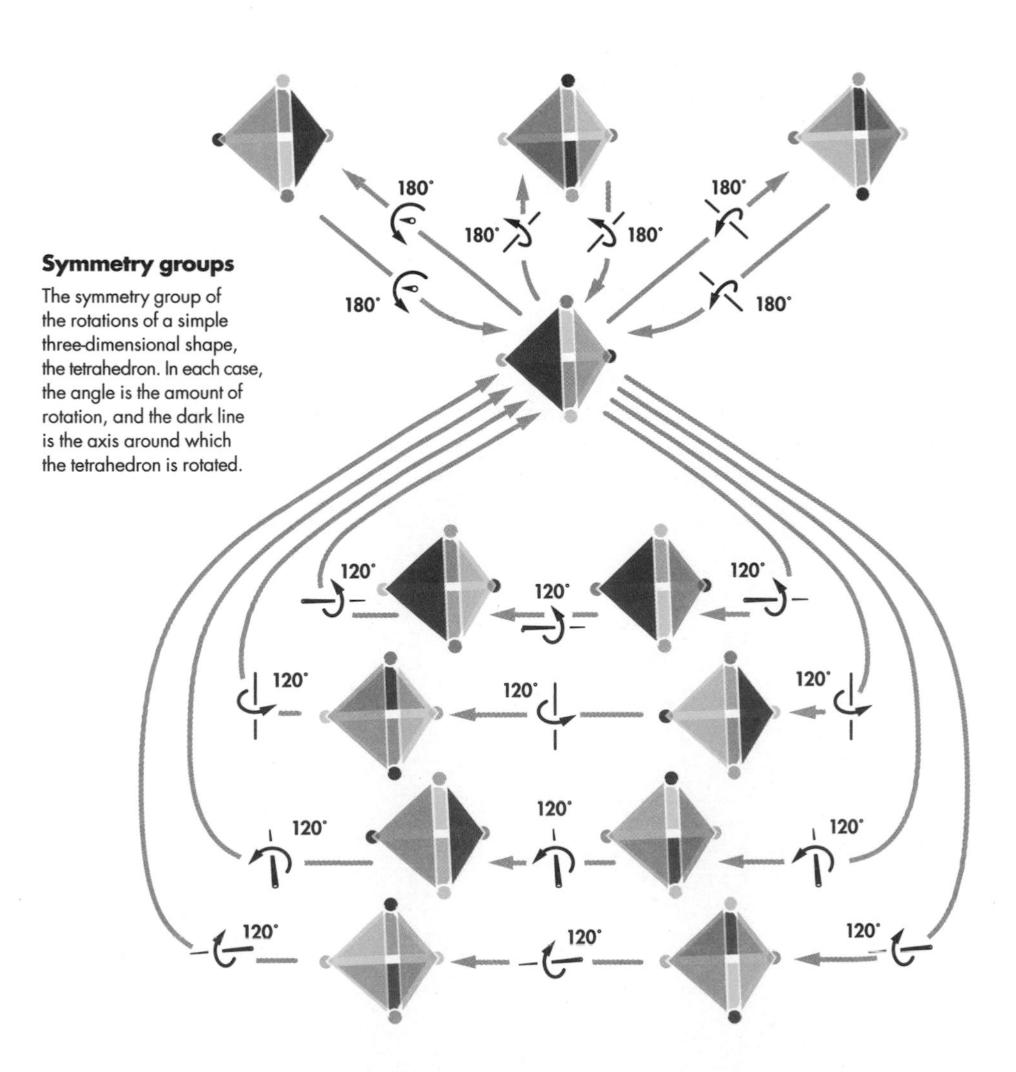

Symmetry groups

The symmetry group of the rotations of a simple three-dimensional shape, the tetrahedron. In each case, the angle is the amount of rotation, and the dark line is the axis around which the tetrahedron is rotated.

SYMMETRY TAKES OVER PHYSICS

One of the greatest contributors to the understanding of symmetry, of whom many people may never have heard, was the German mathematician Emmy Noether. In 1915, Noether showed that symmetry was not only useful in describing crystals and other physical structures, but that it lay behind the conservation laws of physics. These physical laws describe how certain quantities, such as the amount of energy in a system, cannot be created or destroyed. They will always be conserved, although they may be changed into a different form.

As a woman in a man's world, Noether did not have it easy. She was only the second woman ever to gain a PhD in mathematics in Germany, and despite her obvious genius, the system initially prevented her from taking up the option of "habilitation," a German requirement to become a professor. It was only as a result of two special petitions to the government that this was allowed, and even then she still was not awarded a professorship. Noether had Jewish ancestry and also had Communist sympathies, which made her a target of the Nazi regime. She moved to America in 1933, but died two years later at the age of 53.

Noether's greatest contribution was to prove mathematically that each of the conservation laws, such as the conservation of energy or momentum, is linked directly to a symmetry in nature. If those symmetries did not exist, neither would the conservation law. In doing so, she invoked a kind of symmetry that we haven't yet

Spontaneous symmetry breaking

A pencil standing on its point is symmetrical, but the tiniest influence will produce a broken symmetry with the pencil pointing in a particular direction.

met—symmetry in time. Although this is harder to envisage with the visual approach to symmetry, it makes perfect sense in the mathematical invocation of the subject. Something symmetrical in time is unchanged as a result of shifting it through time, which we usually assume to be the case with physical laws.

Noether proved that if the physical laws are indeed symmetrical in time, then energy will be conserved in a closed system. (And conversely, if energy is conserved, the laws must by symmetrical in time.) Translational symmetry of the natural laws—that they apply the same way in a different location—results in the conservation of linear momentum, while rotational symmetry gives us the conservation of angular momentum.

This was only the start of the role of symmetry in physics. As the nature of the forces of physics became better understood, applying the mathematics of symmetry made it possible to contemplate that at the very beginnings of the universe at least some of the forces of nature had been combined, and that early on they split off, due to a process known as "spontaneous symmetry breaking." This is often described as being like the result of balancing a pencil on its point. In principle it is possible to do so, but the slightest movement of air or vibration will result in the pencil falling over in an unpredictable direction.

This isn't a perfect description of the spontaneous symmetry breaking, proposed for the emergence of the forces, as the pencil's fall isn't truly spontaneous—it needs something, however small, to give it a push—but it illustrates the principle. In the early universe, quantum effects that effectively throw in random variations are expected to have been dominant, making such quasi-spontaneous symmetry breaking possible.

Symmetry considerations would also enable physicists to understand how the fundamental particles that make up matter are structured. This began with the observation of the similarity, other than their electrical charge, of the two kinds of heavy particles that make up the atomic nucleus—the proton and the neutron. The German physicist Werner Heisenberg proposed that there was a kind of symmetry between the two types of particle, which he called isospin (a slightly confusing name as it has nothing to do with turning around).

Understanding how a rainbow is formed does not stop us appreciating its beauty. Similarly, understanding the patterns of nature help us to value it more.

PARTICLES AND SYMMETRY

Other physicists, notably the American Murray Gell-Mann and the Swiss George Zweig, also explored the nature of symmetry in these particles. They noticed that there seemed to be an underlying mathematical structure, when introducing another symmetry dimension called strangeness, that produced a pattern of eight positions, characteristic of the symmetry group SU(3). This would lead to the concept of quarks, the fundamental particles that make up protons and neutrons and the "color" charge that quarks and the gluons that hold them together have, which follow this eight-fold symmetry.

Deductions from symmetry would also be behind the search for the Higgs boson mentioned in the particle trail patterns section (see page 71). However, symmetry has not proved a universal panacea for physics. Some scientists have pushed the use of symmetry beyond anything that has been observed to a concept called supersymmetry, which predicts that every particle we know about should have a symmetric equivalent of a different kind. However, these particles have never been observed.

Nevertheless, the patterns of symmetry are fundamental to physics, just as much as they contribute to an object being pleasing to the eye. Symmetry runs through the natural world, and an appreciation of symmetry—and symmetry breaking—helps us both understand it and enjoy it more.

As has been the case with all the patterns we have explored in this book, symmetry is both of interest to the eye and to the mind. Together, these patterns aid our understanding and can be enjoyed in their own right. The poet John Keats accused Isaac Newton of "unweaving the rainbow" because he felt that Newton's science reduced the beauty of the rainbow to mathematics. However, the patterns of nature in reality allow us to both enjoy beauty and appreciate what is happening better. They have a double value that will always enable us to go beyond surface details to a deeper understanding.

"THE THEORY OF ELEMENTARY PARTICLES AND THEIR INTERACTIONS CAN IN ESSENTIAL RESPECTS BE REDUCED TO ABSTRACT SYMMETRIES.

K. V. LAURIKAINEN

INDEX

E

F

H

I

J

INDEX

L

N

P

R

S

T

U

ACKNOWLEDGMENTS

This book is dedicated to Gillian, Rebecca, and Chelsea, and to my late father Leonard Clegg who, as a chemist, sparked my interest in the patterns of science and the way that it helps us to understand the world around us.

Thanks for all the assistance in putting together the pattern that make up the structure of this book, notably from editors Kate Duffy and Kate Shanahan.

Brian Clegg

The publishers would like to thank Wayne Blades for the elegant design and Richard Palmer for his illustrations.

PICTURE CREDITS

The publishers would like to thank the following sources for providing the images featured in this book.

Alamy Stock Photos: 10–11, 21, 24, 25, 67, 184 Science History Images; 46 The History Collection; 55 NG Images; 60 Mark Garlick/Science Photo Library; 102, 212 Phil Degginger; 103 Encyclopedia Britannica/Universal Images Group North America LLC; 129 Ryan McGinnis; 140 Emanuel Lattes; 168 Sabena Jane Blackbird; 187 Axel Kock; 196 Axeley Kotelnikov; 200 Ian Dagnall Computing; 209 Andrey Mihaylov; 213 History and Art Collection.

Getty Images: 96 SSPL.

iStock: 161 Sarah Hamilton; 164–5 cynoclub; 178–9 tampatra; 186 SilverV; 195 tracielouise; 206 boule13.

NASA: 19; 115 Nilfanion; 38 ESA, J. Hester and A. Loll (Arizona State University).

Science Photo Library: 13 NASA; 54 David Parker; 130–1 NASA/Jesse Allen, Earth Observatory/Modis Land Group; 159 Peter Chadwick; 181 Will & Deni McIntyre.

Shutterstock: 4–5, 198–9 Elfinadesign; 17 Suriya KK; 52–3 arleksey; 62: Swen Stroop; 68 danm12; 70 sakkmesterke; 72–3 D-VISIONS; 73 Master Andrii; 80 Africa Studio; 87 Lamyai; 97 Jason Winter; 108 Gilmanshin; 112–13 briddy; 117 Vladi333; 118, 120 Rainer Lesniewski; 121 Mathias Berlin;122 jasminlovesTheOcean; 122 elRoce; 132 jon sullivan; 134–5 KMNPhoto; 137 Min C. Chiu; 144 Laborant; 156 Everett Collection; 158 Alex Smyntnya; 162 Ryan M. Bolton; 170 GUDKOV ANDREY; 201 tr3gin; 203 Bruno Ismael Silva Alves; 204 FX; 207 Zsschreiner; 210 pedrosala.

8–9 Courtesy of @michael75/Unsplash.
12 Courtesy of Bell Labs/Nokia. Reused with permission of Nokia Corporation and AT&T Archives.
78 Courtesy of the Archives, California Institute of Technology. 93 Courtesy of Jacob Bourjaily.
154–5 Courtesy of www.onezoom.org.

Wikimedia Commons Images
28–9 Adam Evans (CC by 2.0); 32 [PD-US-Expired]; 34 Lucien Chavan. Cropped from original at the The Albert Einstein Archives, The Hebrew University of Jerusalem; 65 Anderson, Carl D. (1933). "The Positive Electron". *Physical Review* 43 (6): 491–4. DOI:10.1103/PhysRev.43.491 (Public Domain); 76 California Institute of Technology. Professor Richard Feynman, 1986. Source The Big T (yearbook of the California Institute of Technology) (Public Domain); 111 Rezmason (CC BY-SA 4.0); 114 Vilhelm Bjerknes (CC BY-SA 4.0); 136 Georg Cantor; 146 Cayley Q8 quaternion_multiplication graph.svg; 157 PLOS Biology (CC BY); 160 Darwin's finches by Gould 1; 166 Nick Hobgood (CC BY-SA 3.0); 173 Charles Darwin's 1837 sketch; 174 [PD-US-expired] (Public Domain); 175 Haeckel's original (1866) conception of the three kingdoms of life, including the new kingdom Protista; 177 Ivica Letunic: Iletunic. Retraced by Mariana Ruiz Villarreal: LadyofHats; 180 Marjorie McCarty (CC BY 2.5); 183 (bottom) National Human Genome Research Institute; 188 MM. P. J. Smit & J. Green/[PD-US-expired] (Public Domain); 189 JJ Harrison www.jjharrison.com.au (CC BY 3.0); 193 Augustus Binu (CC BY-SA 3.0); 194 Christoph Bock (CC BY-SA 3.0); 197 Centers for Disease Control and Prevention's Public Health Image Library/Dr. Fred Murphy; 215 Debivort. Tetrahedral group 2.svg; 218 Captain76 (CC BY-SA 3.0).

Illustrations by Richard Palmer: 14–15, 22, 26, 37, 48–9, 82,127, 146 (after Cayley), 215 (after Debivort), 216.

Every effort has been made to trace copyright holders and acknowledge the images.
The publishers welcome further information regarding any unintentional omissions.